Fundamental of Biodiversity Taxonomy

Fundamental of Biodiversity Taxonomy

Jaspal Juneja

CYBER TECH PUBLICATIONS
4264/3, Ansari Road, Daryaganj, New Delhi-110002 (India)
Ph.: 011-23244078, 011-43559448 Fax: 011-23280028
E-Mail: cyberpublicationsdelhi@yahoo.com
Website: www.cybertechpublications.com

FUNDAMENTAL OF BIODIVERSITY TAXONOMY

Jaspal Juneja

First Edition 2010

Published by :

G.S. Rawat for **Cyber Tech Publications**
4264/3, Ansari Road, Daryaganj, New Delhi-110002 (India)
Ph.: 011-23244078, 011-43559448 Fax: 011-23280028
E-Mail: cyberpublicationsdelhi@yahoo.com
Website: www.cybertechpublications.com

Printed at: Asian Offset Printers, Delhi-53

Preface

The depletion of biodiversity is an alarming problem all over the country. The world conservation strategy suggests that the initial effort of biodiversity conservation should aim at establishment and maintenance of a network of protected area systems by making policy changes involving local people in the protected areas management and mobilising financial resources for their conservation and protection. The problem of biodiversity conservation has become a global issue. It is being realised that forests existing in a country is not a resource just for that country, but for the whole of the world.

The Amazoian Rain Forests have been called the "Lungs of the World" as they serve to purity of the global atmosphere by release of oxygen and absorption of pollutants. The rate of deforestation is several times higher in the developing countries than the developed countries, as the forests are being felled to generate funds and space for development.

The total number of species in the world is estimated to be around 5 to 30 million but of which about 1.4 millions species have been described. The total number of plant species in India is estimated to be about 45, 000 (15, 000 flowering plants, 64 gymnosperms, 2843 bryopthes, 1042 pteriodophytes, 1940 lichans and 23000 fungi). Nearly 4900 of those species are endemic to India out of which 1500 are highly threatened.

We know that there are tens of millions of plant and animal species, but we do not know enough to be able to describe the patterns and processes that characterise the distribution of species in space, time and taxonomic groups. Given that in practical terms it is impossible to expect to be able to document biodiversity with any degree of completeness other approaches must be used. Scaling rules offer one possible framework, and this book offers a synthesis of the ways in which scaling theory can be applied to the analysis of biodiversity.

Scaling Biodiversity presents new views on quantitative patterns of the biological diversity on earth and the processes responsible for them. Written by a team of leading experts in ecology who present their most recent and innovative views, readers will be provided with what is the state of art in current ecology and biodiversity science. The book could serve as a starting point for any quantitative biodiversity research.

—*Author*

CONTENTS

1

Introduction

The continued health of human societies depends upon a natural environment that is productive and contains a wide diversity of plant, animal, and microbe species. Life on the earth comprises at least 10 million species of plants, animals, and microbes, while in the United States there are an estimated 750,000 species, of which small organisms such as anthropods and microbes comprise 95 percent.

The sustainability of the forest ecosystems and other natural ecosystems are in danger from the expanding world population, which now totals more than 6 billion. With an estimated growth rate of 1.4 percent per year, it is projected to reach 12 billion by the year 2050. Further, due in large part to the growing human population and diverse human activities (supported in large part by fossil fuels), the current extinction rate of species ranges from approximately 1,000 to 10,000 times higher than natural extinction rates. This is alarming for several reasons. Foremost, biodiversity is essential for the sustainable functioning of agricultural, forest, and natural ecosystems upon which human survival and health depends. The loss of a key species (e.g., loss of a predator) creates an imbalance among the remaining species, and can sometimes result in the collapse of the entire ecosystem. Altering a habitat may also improve the environment for an infectious disease, like dengue.

Species diversity affects the quantity and quality of human food supply. For example, conserving pollinators and natural enemies of pests is essential for successful grain, fruit, and vegetable production. Improving food production decreases malnutrition. Yet, at present, the rapidly expanding human population is intensifying the need for increased food supplies. In the year 2000, more than 3 billion people were suffering from malnutrition—the largest number and proportion of people to date. Each year, between 6 million and 14 million people die from the effects of malnutrition.

In many parts of the world, especially in developing countries (e.g., in the Sahelian region of Africa), severe shortages of vitamin A are causing blindness and even death. Worldwide, approximately 250 million children are vitamin A deficient, and each year vitamin A deficiency causes approximately 2 million deaths and 3 million serious eye problems, including blindness.

Similarly, iron intake per person has been declining, especially in sub-Saharan Africa, South Asia, the People's Republic of China, and South America, because overall shortages of food result in inadequate nutrition. In 1998 more that 2 billion persons were sufficiently iron deficient to cause anemia in 1.2 billion people. An estimated 20 percent of the malnutrition deaths are attributed to severe anemia.

Malnutrition is also associated with parasitic infections that are found in areas were conditions of poverty and inadequate sanitation also exist. The health of malnourished individuals, especially children, is seriously affected by parasitic infections, because their presence reduces the availability of nutrients. Parasitic infections diminish appetites while increasing the loss of nutrients by causing diarrhea and dysentery. Hookworms, for instance, can suck as much as 30 milliliters of blood from an infected individual each day, lowering his or her resistance to other diseases. Because an estimated 5 to 20 percent of an individual's daily food intake is used by the body to offset the effects of parasitic illnesses, the overall nutritional status of a parasite-infected person is greatly diminished over time.

As a human population continues to expand and biodiversity declines, waste grows and its disposal becomes a major environmental problem. Each year the total quantity of waste produced by humans, livestock, and crops weighs about 38 billion tons worldwide. Numerous invertebrate animals and microbes function to degrade and recycle wastes. Their preservation in ecosystems is essential to maintain a healthy and productive environment.

Worldwide chemical waste and pollution are also major environmental problems. In the twenty-first century in the United States, 80,000 different chemicals are used and released into the soil, water, and air; worldwide, an estimated 100,000 chemicals are used. In the United States, more than 1,100 kilograms of chemicals per person are used each year; nearly 10 percent of these are known carcinogens. Each year nearly 3 billion kilograms of pesticides are applied worldwide. These toxic chemicals cause 26 million human poisonings annually, with about 220,000 deaths, and affect approximately 750,000 people with chronic diseases like cancer.

Approximately 75 percent by weight of the chemicals released into the environment can be degraded by biological organisms. Thus, species biodiversity helps provide continuous cleanup of contaminated sites (such as residue of pesticides in agriculture), and has a significant advantage over other techniques. Conserving beneficial natural enemies not only controls crop pests but also helps reduce the amount of pesticides applied.

In addition to degrading chemicals, some invertebrate and microbe species also degrade and recycle biological pollutants in water resources. Again, the biological pollution problem is particularly serious in developing nations. About 1.2 billion people in the world lack clean, safe water because most household and industrial wastes are dumped directly into rivers and lakes without treatment. This pollution contributes to the rapidly increasing incidence of diseases in humans. Waterborne infections account for 80 percent of all infectious diseases worldwide and 90 percent of all infectious diseases found in developing countries. A lack of sanitary conditions contributes to about 2 billion human infections of diarrhea, resulting in about 4 million deaths, per year, mostly among infants and young children. Sometimes altering a natural habitat inadvertently leads to the spread of an infectious disease. Diseases like schistosomiasis that are associated with contaminated fresh water are expanding worldwide. In 1999 it was estimated that schistosomiasis caused 1 million deaths per year. The escalation of the incidence of this disease followed an increase in suitable habitats for the snail that serves as the intermediate host of the causative agent, *Schistosoma mansoni*. Thus, construction in 1968 of the Aswan High Dam in Egypt and its related irrigation systems was followed by an explosion in the prevalence of *Schistosoma mansoni*, which increased in the human population from 5 percent in 1968 to 77 percent in 1993.

Considered together, the natural biodiversity of plants, animals, and microbes functions in many ways to enhance the health and quality of life enjoyed by human society. In view of the likely continued growth in human population, and the resultant alteration of the earth's fragile natural ecosystem, greater efforts must be made to conserve biodiversity as a natural and essential treasure.

RATIONALE OF THE THEMATIC NETWORK

Coastal zones include numerous different functions and uses which depend on the same limited resources and space. This fact frequently generates conflicts between different types of use. Furthermore, interventions regarding single functions of a coastal zone usually have important impacts on other uses within same area or on a vaster territorial scale. Impacts which climate change is expected to have on Europe's Coastal zones are going to accentuate existing and create new conflicts among uses.

Sustainable decision making and management in such multifunctional areas thus requires the capacity to consider all different economic and social spheres contemporaneously asks for the integration of different interests into the decision making process.

In this context, evaluation approaches offer tools and methodologies which allow the description, analysis and assessment of coastal systems, using rational and systematic procedures.

Furthermore these approaches include the possibility of confronting interests and values of different users and actors coming from different economic and social areas on the basis of criteria and indicators for establishing hierarchies of values or for translating values into terms which can be understood by all actors. Using evaluation and assessment techniques, information is provided which can support shared decision making confronting interests of different stakeholders and users of the coastal area in a co-ordinated and rational manner.

Within the Encora Network, one specific thematic network has been dedicated to discussing this twofold challenge: putting into evidence the problems arising from different and conflicting interests, pressures and impacts which originate from social and economic activities on coastal areas and discussing strategies for the integration of these conflicting interests into coastal management and decision making processes.

The Thematic Network on Multifunctionality and Evaluation has collected and discussed up to date concepts, approaches, tools and methodologies which can be used as support in complex decision making processes and help coastal zone managers to consider the requests and needs originating from the variety of functions in Coastal zones.

Evaluation, understood as a means of informing decision making, is of crucial importance for introducing multifunctionality aspects into coastal zone management. Albeit their great potential, evaluation techniques and integrated decision making are still scarcely employed in European coastal zone management. It is desirable that the development of tools facilitating the multifunctional challenge represented by Coastal zones should be brought further by a collaboration between scientists and practitioners and their use promoted further among coastal zone managers. Considerable progress can be made at the European level by networking activities to spread existing examples of good practice and research amongst European partners.

RATIONALE FOR THE COASTAL WIKI

Within the ENCORA project it was decided to develop the Coastal WIKI to provide the most effective tool for networking and dissemination of the state of the art. The contributions of theme 1 to this State of the Art are grouped in four sections, dedicated respectively to:

- the definition of basic concepts underlying different evaluation approaches
- the description of different approaches at the basis of evaluation strategies, and of
- tools and methodologies which can be applied to the evaluation in coastal zone management contexts.
- case studies that provide examples of application of various methodologies.

Concepts

The definition of some basic concepts underlying evaluation approaches has been considered useful for the understanding of the proposed approaches and methodologies. These definitions consider the relationship between multifunctionality on the one side, being one of the characteristics of coastal areas which represent a major challenge for sustainable decision making and management, and evaluation on the other side, which can be used as a support for decision making in complex and multifunctional contexts.

A second important definition regards the term value, as evaluation is used to attribute values to different options or to base decision making on rational judgements. In this context, the term economic value represents one of the basic concepts which allow for the adoption of a common measure for most of the values expressed by different actors.

Although economic value represents an important concept for the measurement of the value society attributes to a certain asset, some values may not be adapted to being measured directly in terms of market values, as they refer to objects which normally are not treated on markets. In those cases, the extension of the concept of value beyond the worth expressed by market values becomes relevant, introducing a concept of value which comprises, according to the theory of welfare economics, the benefit individuals derive from economic goods or services which not necessarily corresponds to the market value.

The introduction of concept of individual benefit as a measure of social value allows for the evaluation of goods and services which cannot be exchanged on markets, such as natural resources. One of the most relevant concepts in this context is represented by the existence value which expresses the benefit individuals gain from the existence of goods which are not used directly independent of present or future uses that will be made of these assets, allowing thus to introduce the value of the existence of categories as natural resources biodiversity etc. into monetary evaluations.

Whereas accurate measurement of values will be adequate in some cases of evaluation, not all values can be revealed with a congruent effort. In order to facilitate monitoring and decision making processes of complex contexts, indicators based on data easy to access or to measure are used. A concept which is commonly measured by indicators is represented by sustainability, where aggregated judgements on the performance of many different entities with respect to one option have to be produced, leading to the definition of specific sets of sustainability indicators which allow also for international confrontation of performances.

Some impacts from human activities are not continuous, but may occur only with a certain probability and a certain variance in the entity of damages caused. The concept of Risk analysis introduces the concept of considering and quantifying aspects of uncertainty in order to integrate them into evaluation. The concept of Carrying capacity finally is based on the notion that ecosystems have a limited capacity of coping with environmental stresses. Approaches based on this concept point to a quantification of limits for socio-economic uses with respect to coastal systems, as for instance, tourism.

Approaches

Based on the different concepts of values described, different approaches are applied to their assessment. As market values can easily be revealed and used for cost-benefit analyses those approaches based on market values represent the most common ones among evaluation approaches and have become quite widespread for the assessment of efficiency of an intervention.

Cost benefit analysis, based on monetary measurement, can also be applied on the combined evaluation of values based on market values and non market values. In these cases different approaches can be applied to translate non market – values into monetary terms, in order to include the relative values in the cost-benefit analysis. Those approaches are based on the assumption that market goods, properly analyzed, can reveal preferences consumers have with regards to non – market goods. In this way an approximation of the social value attributed to a protected area can be elicited by a computation of expenses tourist make to visit the area or by the hedonic analysis of housing prices in the surroundings of the park. Whereas in this case as well as in the case of housing prices, the value of natural resources is derived from consumer preferences expressed (stated) on the market, a second category of expressions of consumer's preferences regards the revealed preferences, as for instance the consumer's (hypothetic) willingness to pay for a service or a good elicited with the help of questionnaires. In those cases where monetary values elicited in an indirect way do not represent a suitable scale for measurement, values can be expressed in terms of "preferences", within approaches based on judgements and using techniques which allow for the comparison of multiple different criteria without a common measure. Each of these three principal approaches to evaluation are described in a section.

TOOLS AND METHODOLOGIES

The section tools and methodologies gives an overview on the state of the art in terms of tools for evaluation and assessment to be used in multifunctional coastal contexts. A specific attention has been paid to those tools which are able to integrate between different sectors and give account of different types of values, integrate broader social and economic aspects into project evaluations as in the case of Regional economic accounting methods. The group of tools, based on the use of economic accounting at different spatial levels, is able to assess the economic impact generated by existing or planned activities of by foreseen events on the regional economic structure, basing the analysis on input – output matrixes of goods and services, supply chain analysis or on Computable general equilibrium (CGE) models.

A further group of tools presented refers to approaches based on monetary valuation of non market values, either on the basis of stated Contingent Valuation Method or revealed preferences, as in the case of Hedonic Evaluation Approaches which use the decomposition of composite market values in order to estimate the component related to natural resources, or methods of travel cost analysis which base the estimation of values of natural resources on the expenses made by individuals who make use of them for leisure time activities. For each of these tools a short description is made which helps to understand the use and the possible outcomes of the technique, and further readings are indicated for studies in detail of the applications.

The application of these evaluation techniques may be quite expensive in terms of time and data requirement, as direct investigation with the help of questionnaires is required. An alternative to direct evaluation is represented by tools which facilitate the use of existing valuations to new evaluations. These tools can be used if data is available from sites with similar characteristics to those of the situation to be examined and allow for the transfer of values estimated in one site to another one.

Further tools for evaluation make use of Multicriteria techniques, which allow for an assessment of policy options based not only on cost-benefit or cost-efficiency criteria, but also on non monetary values or on preferences. An important element of these instruments is the choice of criteria, which have to be shared and understood by all actors, using either predefined sets of criteria such as sustainability indicators or developing proper sets of criteria to be applied to the problem or development at the stake in the specific decision making process. Therefore the involvement of stakeholders is essential, as the participation of possibly all relevant actors into the decision making process can provide for sustainable and shared decisions. The application of multicriteria techniques is thus strongly connected to tools for stakeholder analysis, as a preparatory tool in decision making processes. Further applicative tools in this sector are deliberation and decision support tools as well as knowledge mediation tools, which assist in structuring and informing decision making processes.

Information to be used in decision making processes is highly complex because of the variety of sectors to be considered as well as because of the complexity of single phenomena. Integrated System Frameworks help representing the links between causes, effects and reactions, facilitate the understanding of problems and assist and inform the choice of indicators for evaluation and monitoring.

CASE STUDIES

Case studies can provide a valuable source of information on possible applications of tools and approaches and to guide future studies across Europe. The collection of case studies presented gives an illustration of strategies for evaluating impacts from different socio-economic uses in coastal zones, as for instance from urbanization which represents one of the most relevant pressure on Mediterranean coastal zones. The study describes different approaches to evaluation of impacts of urbanization on values of coastal landscapes, on the benefits of property owners and of simple users of coastal zones and concludes with suggestions for policy options apt to tackle this challenge. The case study on impacts from maritime transport presents the main findings of a large-scale contingent valuation survey conducted after the Prestige oil spill onto the north-western Spanish coast, considering mainly passive use value lost due to the oil spill and illustrates practical/methodological aspects of the realization of the survey. The case study on impacts from tourist uses of coastal zones represents results from a study on tourist ports based on regional economic accounting methods, concluding that, considering also the indirect impacts of these activities, they might be less profitable than commonly expected. The case study on salinization problems for agriculture in the Dutch delta gives insights on problem structuring within a participatory decision making process. The case study on the risks from maritime activities in the Belgian part of the North Sea illustrates the procedure of risk assessment based on the risks resulting from maritime transport for the Belgian coast where potential impacts on the natural environment can result from collisions, groundings, and other incidents either among ships or with fixed marine structures such as platforms or wind turbines.

Development and implementation of ICZM in EU coastal nations: present status, success factors and promising strategies – evidence from the evaluation of ICZM in Europe.

As put into evidence by the recent report of the evaluation of Integrated Coastal Zone Management (ICZM) in Europe [1], the integration between aspects of protection of natural resources and different conflicting interests of coastal uses has only partially been achieved. Conflicts between social and economic interests on the one side and ecosystem goals of conservation and re-naturation of coastal areas on the other still remain substantially unresolved in most national contexts. Frequently, economic interests tend to buy out ecological and social goals, including landscape conservation. In cases where a conciliation between conflicts has been achieved in national strategies, their translation into practical management or planning measures results difficult.

Existing management strategies either disregard the aspects of conservation of coastal and marine natural areas delt with in Theme

7 or, if these goals are addressed, fail in addressing relevant economic and social dimensions of coastal areas, and very often goals of economic development succeed in dominating upon aims of nature conservation. The difficulty of developing holistic and integrated approaches is observed as a shortcoming in most of the national strategies analyzed. With regards to single relevant impacts, some improvements have been achieved in terms of cross-sectoral interests: for instance, tourism-driven interest on high bathing water quality has successfully succeeded in translating an environmental goal of clean water into an economic one, contributing to an improvement of water quality and soliciting investments for wastewater treatment in some countries. Nevertheless tourism sector results to be, in the Mediterranean area and in Portugal, the main driver of uncontrolled and apparently uncontrollable urbanization increasing the pressure by socio-economic activities on coastal areas.

With regards to tools to be used for the implementation of the EU Recommendation related to the integration of aspects of multifunctionality in coastal zone management practice, the report presented by the working group on Indicators states that the use of the common assessment framework of indicators is still very fragmentary and that only a few countries and regions have started the data collection on the base of the recommended set of indicators.

FUTURE CHALLENGES FOR INTEGRATION OF MULTIFUNCTIONALITY IN COASTAL ZONES

Conflicts in coastal zones, which at present conditions are already difficult to conceal, will only be exacerbated by the consequences of climate change. Coastal zones along with their activities and services that are provided for society will be interested in the consequences of sea level rise, more frequent storm surges and by general impacts of changes in climate such as heat waves, reduced rainfall with all consequences on water resources etc. The challenge of economic assessments and of policies acting on integration between social, economic and natural resources will lay in the support for decision making for adaptation strategies, confronting social, economic and ecologic advantages of alternative adaptation tools, and putting into relationship economic efforts made in terms of mitigation and adaptation with benefits from reduced vulnerability and increased resilience.

The discrepancy between a set of tools which facilitate the integration of different sectoral aspects and the still predominantly sector- oriented management practices in European coastal zones represents a challenge to be considered for future action, although the problem will finding a solution not exclusively on a technical level of new or improved tools but first of all in the field of proactive policies. A challenge of a different character is represented by the need of adaptation to impacts and the consideration of risks that will be generated by climate change on European coastal zones which is expected to affect generally the opportunities and threats for economic activities in coastal zones. In this context, the evaluation of risks generated by future changes will become an integrated component of future planning activities in coastal zones.

CONCLUSIONS

Although throughout recognized as an important issue, integration of aspects deriving from socio-economic pressures into coastal zone management processes is still at its beginning. Strategies and instruments for facilitating integration exist to some extend, but, although approved at a scientific level, they are scarcely used at a day-to-day practice in coastal zone management. The lack of integration of economic and social factors into management strategies and decision making processes seriously hampers good practice in coastal zone management, as conflicts between uses are shifted to the implementation phase, with the risk of strategies and decisions being altered in an uncontrolled manner.

THE BIODIVERSITY ACTION PLAN

This document is a Biodiversity Action Plan (BAP) for the Scottish trunk road network. It sets out the commitment of the Scottish Executive to conserve and enhance biodiversity on Scotland's trunk road network. It is designed to provide information and guidance both for those directly involved in managing the trunk road estate, and for those who work to encourage biodiversity in the region. The Plan seeks to integrate local and national biodiversity initiatives, to ensure consistent and positive effort towards improving the natural environment in Scotland. The overall goal of the Trunk Road Biodiversity Action Plan (TRBAP) is to contribute to biodiversity objectives by protecting Scotland's natural heritage on and adjacent to the trunk road network using cost-effective techniques.

To achieve this goal, a seven point plan has been set down:

1. Work to identify and understand the variety of species and habitats that exist along the network, and the inter-relationships between them and the activities of the trunk road Division.
2. Develop sustainable, cost-effective management practices that work with natural processes.
3. Formulate best practice guidance on maintaining and enhancing areas of value, setting targets for action which are compatible with local, regional and national biodiversity objectives.
4. Work in partnership with others. The active involvement of many different organisations is an essential part of implementing national and local biodiversity objectives. Habitats and species do not respect man-made boundaries, and this Action Plan will take full advantage of opportunities to encourage biodiversity by looking beyond the boundaries of the trunk road estate and working in partnership with adjoining landowners and other organisations.
5. Work to promote biodiversity awareness and an understanding of locally important species and habitats to a wide audience.
6. Influence, through the trunk road Division's Environmental Management System, those third parties and indirect activities which may have an impact upon biodiversity.

7. Develop and publicise progress and examples of best practice, thereby encouraging others to build on the experience gained.

The Scottish Executive is committed to maintaining a safe, efficient road network. Within this overriding objective there are many opportunities for cost-effective contributions to the aims of biodiversity. This Plan sets out how, when and where these can be achieved, and by whom.

Detailed Area Action Plans will be developed in consultation with the biodiversity partnerships as this document is implemented, building on the outline plans included in Part C. These will set out objectives, targets and actions for locally important habitats and species.

BACKGROUND TO BIODIVERSITY

Biodiversity is short for 'biological diversity', meaning the whole variety of all living things on earth, including plant and animal species and their genetic diversity. The term also includes the way these organisms interact. In short, the term biodiversity means the whole living world, of which we are an integral part.

Human activity is damaging the very environments on which we depend, causing the destruction of habitats and, in extreme cases, the extinction of species. Biodiversity conservation is about safeguarding species and their habitats. It includes the protection of high profile endangered species but also, crucially, locally important wildlife and habitats in all parts of Scotland.

In recognition of the urgent need to safeguard the Earth's biodiversity, the UK signed the Convention on Biological Diversity at the Earth Summit in Rio de Janeiro in 1992.

As a result, the UK government has produced *'Biodiversity: the UK Action Plan'*. This sets out broad strategies for conserving and enhancing wild species and wildlife habitats in the UK for the next 20 years. The UK Biodiversity Steering Group has developed national aims and objectives to implement these strategies.

In Scotland, a Biodiversity Group has been set up as a partnership, linking key organisations in the public and private sector. The group provides information and practical guidance for those involved in preparing Biodiversity Action Plans. It also oversees the production of species Action Plans for plants and animals which are mainly or wholly found in Scotland (such as corncrakes, red squirrels and chequered skipper butterflies). In addition, habitat Action Plans are prepared for Scotland's most special wild places, including pine forests, blanket bogs and the machair of the Western Isles.

LOCAL BIODIVERSITY ACTION PLANS (LBAPS)

Biodiversity is a key factor in making a local area distinctive and recognisable. For example, native pinewoods with their distinctive flora and fauna are an integral and essential element of the landscape of the central and northern Highlands. Here and elsewhere national targets need to be translated into local action to make a difference. Practical action is being taken to protect species and habitats at a local level through the development of Local Biodiversity Action Plans (LBAPs). LBAPs are being produced for all regions across Scotland, and this TRBAP has been developed to support and complement them. A LBAP identifies which species and habitats are important, rare or threatened in an area, and sets out what should be done to protect and enhance them. It describes how this should be done, promotes partnerships for action and raises public awareness about biodiversity in general. LBAPs in Scotland are being developed by networks of organisations such as local authorities, wildlife groups and local businesses.

BIODIVERSITY AND SUSTAINABILITY

Sustainable development has been defined as:

'Development that meets the needs of the present generation without compromising the ability of future generations to meet their own needs"

(Brundtland Report, World Commission on Environment and Development 1987).

Protecting biodiversity is an integral part of achieving sustainability in the long term. There are close links with local authority plans for tracking and promoting the quality of the environment indicators of sustainability, of which biodiversity is one.

THE SCOTTISH EXECUTIVE TRUNK ROAD BIODIVERSITY ACTION PLAN (TRBAP)

The Scottish Executive welcomes the current initiatives on biodiversity, which reflect its long-term commitment to the environment. The Transport and Planning Group has been carrying out Environmental Assessments and implementing environmental mitigation measures for many years. This document is a continuation of that commitment. It will identify targets and objectives which promote and protect biodiversity on the trunk road network, and will integrate with LBAPs and provide a link to other initiatives.

These efforts are supported in a wider policy context.

> *"is likely to involve commitments to...minimise the impact of roads and traffic on the global and more local environment, e.g. by protecting biodiversity, and employing principles of sustainability in the management of trunk roads...".*

SPECIAL ISSUES FOR THE TRUNK ROAD NETWORK

The objectives of this Plan must be compatible with the key responsibilities of the road operators, including any engineering and operational constraints relating to road management.

This Plan is different to a LBAP, principally in that it relates to an operational facility designed to transport users safely and efficiently, rather than to a geographic area. Trunk roads are linear features which are under the control of one organisation, but which cross geographic, administrative and ecological boundaries. The land area covered may encompass many different types of landscape and habitat. This creates opportunities to work in partnership with the many landowners whose land adjoins the trunk road estate, and with the government, environmental and business groups which have formed to address biodiversity throughout Scotland.

THE POTENTIAL INFLUENCE ON BIODIVERSITY IN SCOTLAND

Trunk roads are roads that are considered to be of strategic importance to the wellbeing of the country and are under the authority of the First Minister for Scotland. The trunk roads are operated, managed and maintained by the Scottish Executive Transport and Planning Group on behalf of the Scottish Parliament.

Trunk roads can be motorways, dual carriageways, single carriageways or single track roads. The trunk road estate includes both the road surface and the verge. The Scottish trunk road network comprises over 3400 km of road, which equates to approximately 6% of the total length of public roads in Scotland. Trunk roads carry approximately half of all traffic on major roads in Scotland and cover an area of over 70 square kilometres, of which more than half is vegetated. This large area of soft estate has considerable influence in biodiversity terms because:

- The land area is spread over a wide network passing through many different landscape character areas and habitat types;
- The soft verges and earthworks, which can be relatively wide in places (over 100 m at some junctions on motorways), are often amongst the few places where wildlife can thrive in intensively managed agricultural landscapes.
- At a wider level, the design and management practices adopted by the Scottish Executive on the trunk roads are often regarded by local authorities to be appropriate for the non-trunk road network, which equates to approximately 51,000 km.

This major land holding carries associated responsibilities for its management and use. Much of the estate is rural, although a significant proportion lies in and around towns and cities. The character of trunk road land varies enormously, ranging from semi-natural grassland and woods to formal and ornately planted features such as roundabouts in urban areas. It presents a significant opportunity to contribute positively to the Scottish environment.

Typical activities associated with trunk road management.

Maintenance	Resurfacing Drainage Road Marking Sign Cleaning Grass Cutting Chipping Weedkilling Thinning Pruning
Management	Policy Best Practice Training Traffic Managent Landscape Management
Improvements	Roads Bridges Structures Drainage Earthworks Landscape

In discharging its responsibilities to maintain a safe, efficient road network, the Scottish Executive Transport and Planning Group controls a wide range of activities. It is important that each is assessed in terms of its impact on the environment if we are to ensure that biodiversity objectives are acheived.

The TRBAP can make a positive contribution to biodiversity in Scotland and to each LBAP area, without compromising the primary transport responsibilities of the Scottish Executive. The integrated approach that is being adopted though this Plan will support sustainable improvements in landscapes and habitats. It is hoped that the examples of good practice given here will encourage other land custodians to take positive action and demonstrate the benefits of partnership initiatives.

ISSUES AND OPPORTUNITIES

Understanding the key issues and opportunities associated with the trunk road estate is critical when developing the objectives of the TRBAP. Many of these issues have already been recognised by the trunk road Division, and positive action has been taken in a number of instances.

FRAGMENTATION OF HABITATS

Some disruption of habitats is unavoidable when carrying out any development, including the construction of roads. However, design and positive management can also enable the road network to link fragmented habitats, increasing their value for a range of species.

USE OF CHEMICALS

Control of over-vigorous weeds along roadsides is essential for safety reasons to prevent them becoming a hazard or public nuisance. This normally involves the use of herbicides. Alternative options, such as the selection of slower growing, lower maintenance plants along new verges, can reduce reliance on chemicals,and promote a more valuable and interesting variety of plants and animals.

ACCIDENTAL POLLUTION

Pollution from roads can potentially affect the flora and fauna of verges and adjacent habitats.

For example, accidental spillages during construction and maintenance could affect nearby watercourses. Such problems can be avoided or minimised by best working practices.

CHANGES TO HYDROLOGY/DRAINAGE

Road construction and drainage can alter the natural drainage pattern of the locality. This in turn can affect habitats which in certain cases may be some distance from the road. This type of problem can be minimised by identifying sensitive habitats and taking steps to avoid or mitigate adverse impacts. Road drainage features, such as ditches, can become valuable wildlife habitats if they are managed sensitively.

Disturbance

Busy trunk roads can cause disturbance to certain sensitive species. For example, some bird species may be affected by noise and others by movement. However, different species display a range of tolerances to such disturbance, and valuable habitats can be maintained even next to very busy roads. The design and management of roadside planting can help reduce such effects. Typically, belts of scrub can be used to provide shelter for birds using adjacent wetland habitats.

Road Crossing

Certain species, such as badgers, otters, deer, hedgehogs and amphibians, are at risk from being run over when attempting to cross roads. Larger animals can also present a safety hazard to traffic.

Measures can be taken either to deter animals from crossing at dangerous locations or to help them cross at more suitable points. For example, underpasses or bridges can be provided at locations where badgers regularly cross a busy road.

Setting Objectives

The following general objectives apply to the whole Scottish trunk road network. More specific targets are set out in individual Area Action Plans (Part C of this document).

- Adopt practices which help to maintain and enhance the biological resource.
- Use native tree species wherever they can effectively fulfil the requirements of the landscape design and management strategy. Where native species are to be used, the selection and procurement of plants shall be in accordance with Forestry Commission Guidance and Appendix 1 of *Cost Effective Landscape: Learning from Nature (CEL:LfN).*

- Identify all designated sites (such as Sites of Special Scientific Interest) adjacent to trunk roads, and ensure that the operation and maintenance of the road does not adversely affect them.
- Identify areas adjacent to or along trunk roads which are considered to be of local value. This information may be obtained from several sources, including LBAPs and local nature conservation organisations.
- Encourage the creation of diverse ground conditions where this does not contradict the requirements of the wider route landscape design and management strategy.
- Make repairs to landscapes and habitats which may be unavoidably affected by road improvement or maintenance activities. Seek opportunities to make further improvements to sites where appropriate.
- Identify and describe appropriate protection or enhancement measures that should be implemented as part of maintenance or engineering works (see Part B, the Toolkit).
- Set biodiversity related objectives for managers and staff which reflect local and national targets for important habitats and species.
- Ensure staff are aware of biodiversity objectives and integrate positive biodiversity action into their normal working practices.
- Reduce reliance on fertilisers and herbicides.
- Draw up targets for action across the network which are compatible with local, regional and UK biodiversity objectives.
- Make contact with LBAP Steering Groups or local biodiversity representatives.
- Report on progress and publicise actions taken. This also serves to highlight the positive work of managers and staff throughout the network, without whom this Action Plan could not be implemented.

ACHIEVING OBJECTIVES

The Transport and Planning Group intends to identify areas alongside trunk roads which have particular biodiversity interest, and areas where opportunities for enhancement exist. This task will be carried out in partnership with existing Local Biodiversity Steering Groups, nature conservation networks and other organisations. Joint projects, for instance with adjoining landowners and local groups, can have a beneficial effect on a wider area than the immediate road environment. Stimulating public awareness and commitment to biodiversity objectives is a priority for any BAP. Opportunities exist to use high profile or frequently visited trunk road locations to put across the biodiversity message, for example in urban areas, at lay-bys and by service stations.

Roads are an integral part of leisure and amenity access for people. A large proportion of tourism in Scotland is based on the richness of the country's natural habitats, wildlife and scenery. The Tourism and Environment Task Force, a body established by the Secretary of State, is supporting local partnerships across Scotland to look at issues of sustainable tourism development and to raise project funding. There are therefore potential partnership links with organisations such as the Tourist Board which promote the locality.

IMPLEMENTATION AND REVIEW

Biodiversity should not be an 'add-on' to the work of the trunk road Division, but should be part of its normal business. In time, biodiversity objectives will become fully integrated into existing policies, practices and planning. The long term strategy is to make the protection and promotion of biodiversity as much a part of normal, everyday working good practice as are health and safety.

This BAP will be implemented through four mechanisms. These are, in order of importance and value to biodiversity:

Day to day activities of Trunk Road Authorities

This mechanism has the greatest scope for the protection of biodiversity as it involves consideration of objectives by all personnel in all their activities.

Landscape Action Plans

TRBAP actions will be delivered through the Landscape Action Plan mechanism described in Part B.

ENVIRONMENTAL WORKS FOR ROAD IMPROVEMENT AND MAINTENANCE SCHEMES

The environmental mitigation measures designed and implemented for road projects can contribute significantly to biodiversity.

Implementation Mechanisms for the TRBAP

The objectives and targets of the TRBAP will be built into the contracts of these projects.

SPECIFIC BIODIVERSITY ACTION PROJECTS

It is proposed that a small budget be set aside to carry out minor trunk road related biodiversity projects in partnership with other organisations or landowners.

Review

This BAP is part of an iterative process, not an end in itself. It has therefore been structured to allow for review and amendment in the future.

The TRBAP will be reviewed and updated periodically, taking into account developments at a local and national level, for example the further publication of LBAPs, and experience gained through the implementation of the Plan over the previous years.

2

Biodiversity Management

Biological diversity (or biodiversity) is the diversity of plants, animals and other living organisms in all their forms and levels of organization, and includes the diversity of genes, species and ecosystems, as well as the evolutionary and functional processes that link them.

Developing a biodiversity conservation strategy that is based on a variety of management strategies for individual species is neither feasible nor effective. The impact of forest management practices on many species is unknown and certain practices that benefit some species are often detrimental to others. Recommended instead is the development of an ecosystem management approach that provides suitable habitat conditions for all native species. In this way, habitat diversity is used as a surrogate to maintain biodiversity.

At the same time, however, special efforts may be needed to protect the habitat of species known to be at risk, such as threatened, endangered, or regionally important species. Specific strategies for addressing these species are outlined in the *Managing Identified Wildlife Guidebook*.

Planning to maintain biodiversity should occur at a variety of levels all of which are linked hierarchically: provincial (such as the provincial biodiversity strategy), regional (such as the planning being carried out by the Commission on Resources and the Environment), subregional (such as the planning being carried out through Land

and Resource Management Planning), landscape, and stand. This book applies to two of those levels: landscape and stand.

The biodiversity management approach described here is based on ecological principles and will be refined over time as new knowledge is obtained and management practices evolve. The underlying assumption of this approach is that all native species and ecological processes are more likely to be maintained if managed forests are made to resemble those forests created by the activities of natural disturbance agents such as fire, wind, insects, and disease.

It has been these natural ecological processes, along with burning by aboriginal peoples, that have determined the composition, size, age, and distribution of forest types on the landscape, as well as the structural characteristics of forest stands.

Principles and assumptions

- The more that managed forests resemble the forests that were established from natural disturbances, the greater the probability that all native species and ecological processes will be maintained.
- The habitat needs of most forest and range organisms can be provided for by:
  - maintaining a variety of patch sizes, several stages, and forest stand attributes and structures across a variety of ecosystems and landscapes
  - maintaining connectivity of ecosystems in such a manner as to ensure the continued dispersal and movement of forest- and range-dwelling organisms across the landscape
  - providing forested areas of sufficient size to maintain forest interior habitat conditions and to prevent the formation of excessive edge habitat.
- To sustain genetic and functional diversity, a broad geographic distribution of ecosystems and species must be maintained within forest and range lands.

- Management for biodiversity must be flexible and adaptive. This guidebook provides recommendations rather than specific prescriptions for managing biodiversity. Success in meeting the intent of these recommendations depends on the innovativeness and creativity of land managers.
- Not all elements of biodiversity can be—or need to be—maintained on every hectare. The intent is to maintain in perpetuity all native species across their historic ranges.
- Management for biodiversity should be applied within landscapes regardless of administrative boundaries. Where natural landscapes have been administratively divided, management agencies and licensees should develop a biodiversity plan together. Landscape units are the basis on which the success of biodiversity management will be evaluated.
- The conservation of biodiversity depends on a coordinated strategy that includes:
  - a system of protected areas at the regional scale
  - provision for a variety of habitats at the landscape scale
  - management practices that provide important ecosystem attributes at the stand scale.
- Intensive forestry and other resource development within managed landscapes can be compatible with the maintenance of biological diversity.
- Where past forest management practices have resulted in forest conditions that prevent biodiversity objectives from being achieved, biodiversity can be restored over time by managing the forest to create—or recover—the required ecosystem elements.

RELATIONSHIPS TO OTHER GUIDEBOOKS

This guidebook describes future desired conditions for forests and grasslands at the landscape and stand levels. Other guidebooks that also provide direction on maintaining biological diversity at the landscape level are the *Riparian Management Area Guidebook*, the *Managing Identified Wildlife Guidebook*, and the *Regional Lakeshore Guidebook*. Riparian management areas (RMAs) and wildlife habitat areas (WHAs) can contribute to meeting the old-growth and connectivity objectives within a landscape unit. It is also likely that RMÅs and WHAs will be the main building blocks for the design of Forest Ecosystem Networks (see the section "Designing forest ecosystem networks").

As well as these, several other guidebooks—notably those that address terrain stability, watershed assessment procedures, community watershed management, and visual landscape management—involve land zoning and recommend certain constraints to harvesting and grazing activities. Collectively these guidebooks provide planners with further direction on how land might be zoned into the long-term leave areas needed to minimize the effects on biodiversity of habitat fragmentation and old-growth conversion. The administrative process for establishing, varying or canceling landscape units and landscape unit objectives is described in the *Higher Level Plans Guidebook*.

BIODIVERSITY EMPHASIS OPTIONS

As natural ecosystems become increasingly modified by human activities, natural patterns of biodiversity become increasingly altered and the risk of losing native species increases. The greatest degree of disruption occurs from extreme habitat modifications such as urbanization and agriculture. Parks and protected areas, on the other hand, if appropriately managed, maintain close to natural levels of biodiversity. Managed forest lands fall between those two extremes, and can support varying levels of natural biodiversity, depending on the management practices. More natural levels of biodiversity will be maintained in managed forests if those forests are managed to mimic important characteristics of natural forest conditions.

This guidebook outlines a range of three options for emphasizing biodiversity at the landscape level. Each option is designed to provide a different level of natural biodiversity and a different risk of losing elements of natural biodiversity:

- The lower biodiversity emphasis option may be appropriate for areas where other social and economic demands, such as timber supply, are the primary management objectives. This option will provide habitat for a wide range of native species, but the pattern of natural biodiversity will be significantly altered, and the risk of some native species being unable to survive in the area will be relatively high.
- The intermediate biodiversity emphasis option is a trade-off between biodiversity conservation and timber production. Compared to the lower biodiversity emphasis option, this one will provide more natural levels of biodiversity and a reduced risk of eliminating native species from the area.
- The higher biodiversity emphasis option gives a higher priority to biodiversity conservation but would have the greatest impact on timber harvest. This option is recommended for those areas where biodiversity conservation is a high management priority.

In reality, these options are points on a continuum, and in between lie a range of options that may be selected depending on the relative priority allocated to biodiversity conservation and timber production in an area.

APPLYING BIODIVERSITY EMPHASIS OPTIONS TO LANDSCAPE UNITS

Applying biodiversity emphasis options to landscape units across a subregional planning area is a key part of a biodiversity management strategy.

Biodiversity emphasis options can be assigned to landscape units in the following ways:

- Direct assignment of a biodiversity emphasis option to a landscape unit by a regional or subregional land use planning process.

- Indirect assignment of biodiversity emphasis options to landscape units by a regional or subregional land use planning process that develops broad-scale land use zonation. In such instances agency representatives will assign a specific landscape unit an emphasis option consistent with the zonation.
- In the absence of regional or subregional plans, interim biodiversity emphasis options can be assigned to a landscape unit by joint agreement of the district manager and the designated environmental official.

If no landscape unit has been designated for an area, or an emphasis option has not been assigned for a landscape unit, the default is that the area is managed using the lower biodiversity emphasis option.

To assist with the process of assigning biodiversity emphasis options to landscape units, each unit can be evaluated according to several criteria: topographic and ecosystem complexity, wildlife and fisheries species diversity, significance of key management species and social and economic considerations.

Government has evaluated social and economic impacts against risk to biodiversity on a provincial basis and provided the following policy direction concerning landscape unit biodiversity emphasis assignments within subregional areas. In some cases, large portions of a landscape unit may require a different biodiversity emphasis than the rest of the landscape unit. For example, a low elevation subzone may have a different biodiversity emphasis option assigned than a high elevation subzone. Within a landscape unit, minor areas (up to thousands of hectares in size), may be managed differently from the rest of the unit.

For example, a landscape unit assigned a lower biodiversity emphasis option could contain some small protected areas that would contribute to the seral stage and patch size requirements for the landscape unit. Alternatively, stands managed for intensive timber production may be part of any landscape unit, regardless of the biodiversity emphasis option assigned the major portion. Concentration of the lower biodiversity emphasis option over extensive, contiguous areas of a subzone or subregional planning area (over several adjacent landscape units, for example) should be avoided as it will greatly impact natural levels of biodiversity in that area. Instead, landscape units with the lower biodiversity emphasis should be distributed across the subregional planning area.

ESTABLISHING LANDSCAPE UNIT BIODIVERSITY OBJECTIVES

Once landscape units have been delineated and the biodiversity emphasis option determined, objectives for maintaining biodiversity within the units may be set for the following characteristics:

- several stage distribution
- temporal and spatial distribution of the cut and leave areas "patch size distribution"
- old several retention and representativeness
- landscape connectivity
- stand structure
- species composition.

British Columbia has tremendous ecological variation. Although some general forest and range management practices can broadly accommodate the needs of all ecosystems, more often a variety of practices is needed to respond to the different natural disturbance regimes under which ecosystems have evolved. For the purpose of setting biodiversity objectives, five natural disturbance types (NDTs) are recognized as occurring in British Columbia. They are:

NDT1 – Ecosystems with rare stand-initiating events

NDT2 – Ecosystems with infrequent stand-initiating events

NDT3 – Ecosystems with frequent stand-initiating events

NDT4 – Ecosystems with frequent stand-maintaining fires

NDT5 – Alpine Tundra and Subalpine Parkland ecosystems

These disturbance types characterize areas with different natural disturbance regimes. Stand-initiating disturbances are those processes that largely terminate the existing forest stand and initiate secondary succession in order to produce a new stand. The disturbance agents are mostly wildfires, windstorms and, to a lesser extent, insects and landslides. The stand-maintaining disturbances—such as the understorey surface fires that occur in the Interior Douglas-fir and Ponderosa Pine biogeoclimatic zones—serve to keep successional processes stable.

The following descriptions of the five NDTs include the biogeoclimatic zones, subzones, and variants that fall under each disturbance regime. As additional information becomes available, the number of NDTs and the biogeoclimatic units within them may be revised and refined.

Recommended procedure for establishing landscape unit biodiversity objectives

- Confirm the biodiversity emphasis option for the landscape unit.
- Map the NDTs within the landscape unit, based on the biogeoclimatic subzones and variants present in the unit.
- Use the NDT recommendations to establish the biodiversity objectives for the landscape unit.
- Where a landscape unit consists entirely of one NDT, the biodiversity objectives for the landscape unit should be those of the NDT. However, where a landscape unit consists of more than one disturbance type, biodiversity objectives for the landscape unit should be developed for each of the NDTs present. In areas that are transitional between two NDTs, the management practices should be modified to reflect that transition.
- Where a disturbance type accounts for only a small, isolated part of a landscape unit, meeting the landscape level objectives for the small area may not be possible. In such cases include the small area in an adjacent disturbance type.

ESTABLISHING SEVERAL STAGE OBJECTIVES FOR NATURAL DISTURBANCE TYPES

The composition of plant and animal communities changes as forest stands develop through time after a disturbance. Various organisms find their habitat needs during different stages of forest development (known as several stages, or forest ages) and most specialist species are associated with either the early herb/shrub stage or the mature to old several stages.

Different natural disturbance regimes have created forests with greatly differing several stage distributions. Portions of the province with less frequent stand-initiating disturbance have more older forest, and a greater abundance of species adapted to landscapes of older forests, than do areas with more frequent disturbance.

Forest harvesting generally increases the amount of young forest and decreases the amount of older forest, because commercial forest rotations are generally shorter than natural disturbance return periods. This effect is most pronounced in forest types that have the lowest frequency of natural stand-initiating disturbance. Therefore, the more that managed forests diverge from natural disturbance regimes, the greater the risk of loss of biodiversity.

The seral stage guidelines provided by NDT are intended to help foresters maintain the diversity of seral stages and disturbance regimes found within various biogeoclimatic zones, subzones and variants.

It is unclear to what extent management can deviate from natural seral stage distributions without losing elements of biodiversity. Even at the scale of landscape units, natural patterns often vary from the average.

This guidebook, while acknowledging the uncertainties, presents the minimum requirements considered to have a good probability of maintaining biodiversity within the landscape unit. It assumes that greater change from natural seral stage distributions would increase the risk to biodiversity, and less change would decrease the risk.

CUTBLOCK SIZE AND BIODIVERSITY OBJECTIVES

Section 21 (1) of the Operational Planning Regulations (OPR) states that cutblock design, including size, shape, and pattern, must be consistent with the objectives established under a higher-level plan, including objectives for the maintenance of biodiversity. Recommendations in this guidebook for cutblock maximum size, and pattern are provided in the section on "Temporal and spatial distribution of cut and leave areas." Also included there are recommendations for patch size distribution for cutblocks and leave areas.

The maximum cutblock size that may be proposed is either 40 or 60-ha, depending on which forest region the plan is in (OPR 21 (2)). However, smaller (OPR 21 (3) (a)) or larger (OPR 21 (3) (b)) maximum cutblock sizes may be allowed by the district manager for specific reasons. Where larger cutblocks are specified, the cutblock design must be consistent with the structural characteristics and the temporal and spatial distribution of natural openings (see recommendations in the section on "Stand management to maintain biodiversity").

NATURAL DISTURBANCE TYPE 1: ECOSYSTEMS WITH RARE STAND-INITIATING EVENTS

Historically, these forest ecosystems were usually uneven-aged or multi-storied even-aged, with regeneration occurring in gaps created by the death of individual trees or small patches of trees. When disturbances such as wind, fire, and landslides occurred, they were generally small and resulted in irregular edge configurations and landscape patterns.

The mean return interval for these disturbances is generally 250 years for the CWH and ICH, and 350 years for the ESSF and MH biogeoclimatic zones.

Occasionally, major windthrow events have occurred as a result of hurricane-force winds on certain exposed parts of coastal British Columbia. The average return interval for these has been approximately 100 years. Those portions of the CWHvm1, CWHvm2, CWHvh1 and CWHvh2 variants where these wind events occurred should therefore be considered to fall within NDT3, ecosystems with frequent stand-initiating events.

Biogeoclimatic units in NDT1

The following biogeoclimatic subzones and variants make up this disturbance type:

While stand-initiating fire disturbances also occur—though rarely—in the Interior Douglas-fir and Ponderosa Pine biogeoclimatic zones, these zones are not included in NDT1 because the stand-maintaining surface fires are the dominant ecological influence.

Seral stage distribution (NDT1)

To maintain important landscape characteristics in this disturbance type, a relatively high proportion of forests with mature and old seral stage forest attributes is required. In managed stands this requirement can be met through the use of a combination of silvicultural systems.

Recommendation:

- Seral stages should occur in a variety of patch sizes within the landscape unit.

Temporal and spatial distribution of the cut and leave areas (NDT1)

The pattern and timing of forest harvesting are the dominant factors that determine the size and spatial distribution of similarly aged forest patches in managed landscapes. The objective in this disturbance type is to maintain a range of small to large (up to 250 ha) similarly aged forest patches on the landscape. The forest patch size distribution applies to both harvest units and the leave areas in the landscape units.

Partial cutting and some small clearcuts could be used in this NDT. Complete reliance on small, dispersed clearcuts and small leave areas, however, would lead to excessive forest fragmentation. Therefore, some larger patches should be cut to form larger openings; others should be identified as leave areas.

Recommendations:

- To approximate the historic pattern of this landscape type, a combination of smaller dispersed clearcuts, some dispersed partial cuts, and a few large aggregated harvest units should be used, along with mature and old seral stage forests maintained in a connected network.
- Patch sizes greater than 40 or 60 ha can be created by harvesting the entire larger patch at one time or by aggregating small cutblocks over time. In either case, structural attributes (i.e., live and dead trees) consistent with the natural disturbance type are to be retained within the patch. If smaller cutblocks are aggregated over time, the district manager may waive or reduce green-up requirements to accomplish this. When approved cutblocks exceeding 40 or 60 ha are advertised, and appear in the *Gazette*, the fact that they meet biodiversity retention objectives should be indicated.
- The size range of leave areas should be the same as that for openings.

Old seral stage retention and representativeness (NDT1)

The target for old seral stage retention in this disturbance type is described in the recommendations below.

Recommendations:

- Rare site series should be retained in greater proportion than they occur in the landscape unit; other site series should generally be retained in proportion to their occurrence in the landscape unit. Where site series mapping is not available, a combination of forest cover and site productivity or site index information should be used to determine representativeness.

- The old seral stage retention objective should include patches designed to provide some forest interior conditions across a landscape unit. Where a lower biodiversity emphasis option is chosen, the target for forest interior conditions may be as low as 25% of the old seral area. If an intermediate or higher biodiversity emphasis option is chosen, the target should be 50% of the area. Objectives for obtaining forest interior conditions can sometimes be accomplished most efficiently by increasing old seral stage retention around wildlife habitat areas, riparian management areas, or other suitable areas.
  - For a higher or intermediate biodiversity emphasis area, all the existing old seral forest should be retained, and additional areas designated to be left to become old seral forest and make up the shortfall in the future.
  - For a lower biodiversity emphasis area, the economic and social consequences of halting the timber harvest of old seral forest may be politically unacceptable. If so, some additional harvesting of old seral stands may proceed, and the area equivalent to the shortfall in old seral area must be recruited over time, according to an approved long-term recruitment plan. The old seral retention objective must be in place by the end of three rotations.

Landscape connectivity (NDT1)

Historically, these forests existed as contiguous tracts of old seral stage forest in which stand structure was complex because major stand-initiating events were rare. Current forest practices greatly alter these forests compared to historical conditions.

Connectivity can be maintained through the delineation of forest ecosystem networks. It can also be achieved at a broader scale within landscape units, according to the recommendations under "Temporal and spatial distribution of the cut and leave areas," above. The methods selected should depend on the connectivity objectives of the landscape unit.

Management to reduce fragmentation and maintain connectivity in managed forest landscapes should be guided by the type and degree of connectivity found in each disturbance type. Connectivity can be maintained by a combination of the following methods:

- Maintain late successional forest linkages as part of forest ecosystem networks (see the section "Designing forest ecosystem networks"). These linkages should mimic the type and degree of connectivity found in each disturbance type.
- Manage the forest matrix (the forest between the defined linkages) to maintain additional connectivity through the use of:
  - seral stage and patch size guidelines
  - stand remnants left during harvesting operations (such as small unharvested patches or advanced regeneration)
  - stand management to maintain some of the structural attributes of older forests.

Recommendations:

- Connectivity should be maintained or provided for—especially in those areas identified as "high"—using variable width linkages as part of forest ecosystem networks.
- Tailor the application of seral stage, patch size, and stand structure recommendations to manage the area outside of defined linkages to meet the specific connectivity objectives of the landscape unit.

Stand structure (NDT1)

Maintaining a variety of canopy layers (vertical structure) and spatial patchiness (horizontal structure) is important for maintaining biodiversity in this NDT.

Recommendations:

- Stand management activities should maintain the natural variety of canopy layers and gaps among and within stands across the landscape.

- Even-aged or partial cutting silvicultural systems may be appropriate, but retention of mature forest attributes is especially important in this NDT. These attributes can be retained throughout the stand and / or concentrated within wildlife tree patches (see the section on "Wildlife trees"). Note, however, that these wildlife tree patches should not be additive to those recommended in "Stand management to maintain biodiversity."
- Practices for maintaining stand structure should be considered for all stands and applied as outlined in "Stand management to maintain biodiversity."

Species composition (NDT1)

The section on "Stand management to maintain biodiversity" recommends stand level practices for maintaining species composition.

Recommendations:

- A significant component of the landscape unit should be maintained in communities with plant species composition similar to that in communities that have developed through natural succession.
- Extensive conversion from climax to young seral species, or from young seral to climax or non-native species, should be avoided.
- Rare forest stand types (that is, those accounting for less than 2% of the area, such as birch, cottonwood, aspen, alder and maple) within the landscape unit should be maintained over the rotation.
- The proportion and distribution of the deciduous broadleaf components of stands should be maintained within the range found in unmanaged stands within the landscape unit.

DESIGNING FOREST ECOSYSTEM NETWORKS

This section provides planners with direction in designing landscape units that meet old-growth and connectivity objectives while minimizing negative impacts on timber and range resource values. The recommended long-term landscape management approach is that of creating forest ecosystem networks (FENs) within the landscape units.

A FEN is a contiguous network of representative old-growth and mature forests (some of which provide forest interior habitat conditions) delineated in a managed landscape. Not only does a FEN aim to meet the needs of native species and ecological processes, it also serves to maintain or restore the natural connectivity within a landscape unit.

FENs are composed of a variety of protected areas, as well as classified areas (such as riparian management areas and wildlife habitat areas), any other areas that have been designated as sensitive, high visual quality areas, and areas of unstable terrain or other conditions that make them inoperable.

Most landscapes already contain significant areas within which timber harvesting is constrained. FENs should therefore be designed to take advantage of these areas, using them to form the network "building blocks." This approach makes efficient use of the land base by allowing existing areas with harvesting constraints to serve the dual purpose of protecting a particular resource value while also contributing to the overall structure and composition of the FEN.

In mountain and valley ecosystems with wet climates, where contiguous old-growth forest was a dominant component of the natural landscapes, the delineation of FENs is especially important. However, FENs may also be used in other ecosystems to link important habitats such as wetlands. At the same time, not all components need to be connected and, overall, the need for connectivity varies among disturbance types.

Forest ecosystem networks provide the following benefits:

- They reduce the impact on landscape units of habitat fragmentation and old-growth conversion.
- They represent the full range of ecosystems in the landscape unit.

- They provide some forest interior habitat within each landscape unit.
- They provide wildlife species with areas of refuge during periods of disturbance on nearby sites, as well as acting as centers and corridors of dispersal for the re-colonization of historic ranges by certain species.
- They provide a continuum of relatively undisturbed habitat for indigenous species that depend on mature and old-growth forests.
- They provide daily and seasonal movement corridors for certain species.

Several general considerations in FEN design are:

- Some components of a FEN should be permanent reserves (for example, unstable slopes); others should be sensitive areas that retain important stand attributes (for example, riparian management areas).
- Riparian habitats (found adjacent to streams, rivers, lakes, and wetlands) provide many of the features necessary to maintain biodiversity at the landscape level. In many instances these should form the focal point in FEN delineation. Their linear nature provides species with movement opportunities between ranges at different altitudes, and their diverse vegetation provides species with the structural and functional attributes they need to be sustained. Nevertheless, while riparian areas are important to most species, FENs should not be composed entirely of riparian habitats.
- Where areas previously constrained (such as wildlife habitat areas, riparian management areas, areas with visual quality objectives) are used to meet old seral requirements, they will henceforth be managed as old-growth management areas.
- It is important that all ecosystems in a landscape unit be represented in the FEN designed for that unit. This means that upland habitats such as those on south aspect slopes

and ridge tops, as well as habitats on cooler and moister northerly aspects, should be considered.

- A key component of FENs is, as the name implies, the requirement for important habitat features to be connected in a manner that forms a comprehensive landscape network. Ideally, this connectivity should be dominated by old-growth or mature timber and should be established to incorporate natural terrain features such as gullies and ridges. As shown in each of the five disturbance type summaries, the characteristics of natural connectivity vary by NDT (see "Landscape connectivity" under each disturbance type summary).
- In designing FENs for any one landscape unit, planners should remember to consider the habitat conditions and management plans in adjacent landscape units, as these may affect issues of connectivity and age class distribution in the unit being designed.

Principal steps in designing a FEN

1. dentify and map representative and rare old-growth stands:
   - o Establish a list of representative and rare ecosystems using site series or other mappable surrogates such as forest cover and soils/terrain maps.
   - o Map the location and extent of rare and representative ecosystems.
   - o Assess which forested habitats are of sufficient size to provide forest interior stand conditions.
2. Identify and map the forest areas that will not be harvested. Determine the extent to which they can meet old-growth retention objectives for rare and representative ecosystems.

   A. Existing protected areas

   B. Ecological reserves and wilderness areas

C. Areas subject to harvesting constraints:
   - riparian management areas
   - lake shore management areas
   - wildlife habitat areas
   - terrain sensitive areas
   - recreation areas
   - visually sensitive areas.

D. Inoperable areas

E. Areas that provide habitat for threatened or endangered species (contact the Conservation Data Centre for information)

3. Assess the complex of areas identified above and list representative and rare ecosystems not protected as potential old-growth management areas.
4. Where old-growth retention or representativeness objectives have not been achieved, create old-growth management areas (OGMAs) between the mapped areas.
   - o Ensure that OGMAs are composed mainly of old-growth or mature forest. Use existing forests as models when considering the appropriate appearances of OGMAs.
   - o Where second-growth forests must be used, consider applying silvicultural interventions to develop and enhance mature forest attributes (see Appendix 5). The objective is to provide these attributes in levels that meet the natural levels found on similar sites. Site productivity may be a factor influencing the ability of a stand to provide and sustain mature and old-growth attributes more quickly.
   - o Distribute OGMAs among the low-, mid-, and high-elevation areas within each landscape.
   - o Narrow OGMAs (less than 600 m) may include small linear landscape features such as gullies, ridges, or streams.

Wide OGMAs (greater than 600 m) will provide forest interior conditions. Consider delineating wide OGMAs if such conditions are otherwise absent from the FEN.

- o Consider using riparian areas for a portion of an OGMA.
- o If there are relatively small gaps between mapped areas within a FEN, fill these in with OGMAs to aggregate the gaps into larger units.
- o Recognize that not all areas of harvesting constraint need to be linked.

5. Assess whether connectivity objectives have been achieved. If connectivity objectives were not achieved through retention or representation, use mature and old-growth forest in OGMAs to link stands.
6. Review the resultant FEN with respect to the biodiversity objectives for the landscape unit and check that, within the FEN:
   - o all old-growth ecosystems, as represented by site series or a surrogate, are represented at the minimum level specified in the appropriate seral stage tables in the section on "Establishing landscape unit biodiversity objectives"
   - o rare ecosystems are over-represented
   - o forest interior condition requirements are met
   - o connectivity needs are met.

Old-growth management areas

- OGMAs are intended to capture old growth within landscape units to meet retention objectives.
- Timber harvesting and silvicultural practices within an OGMA should be consistent with management objectives for the OGMA.
- OGMAs can be harvested when equivalent old seral stage areas are available.

- Replacement OGMAs can be brought on stream earlier than would naturally occur through silvicultural interventions designed to promote the key attributes, or through the retention of these attributes during harvesting.

Management within FENs

Recommendations:

- Roads through protected areas, wildlife habitat areas, and sensitive areas within FENs should be avoided.
- The number, length and width of rights-of-way for roads through FENs should be minimized.
- Prompt and appropriate steps should be taken to deactivate roads no longer in use in FENs.
- Where mature and old seral areas (but not areas designated as OGMAs) are identified to meet connectivity objectives, and provided the connectivity objectives can be maintained, some harvesting can occur within these linkage areas as long as the mature stand attributes are maintained.
- When natural disturbances such as wildfire, wind throw, or insect outbreak occur within, or threaten to enter, a FEN, the appropriate management action should be based on an evaluation of the disturbance's effect on the functioning of the FEN. Conversely, when a natural disturbance threatens to affect areas outside the FEN, the appropriate management action should be based on an evaluation of the impact on the adjacent commercial forest or leave areas.
- Where natural disturbances have affected a FEN, management actions such as salvage logging and site rehabilitation must be evaluated (to determine, for example, the value of the FEN and the value of the damaged timber) to ensure that such decisions do not compromise the integrity of the FEN, adjacent commercial forests, leave areas, or other forest values.

STAND MANAGEMENT TO MAINTAIN BIODIVERSITY

This chapter describes how specific objectives for maintaining stand structure, tree and vegetation species composition, and coarse woody debris can be determined.

The recommendations in this chapter should be applied in the preparation of:

- forest development plans
- silviculture prescriptions and stand management prescriptions
- logging plans, fire management plans, and range use plans.

Objectives for wildlife trees and coarse woody debris must be included in the contents of a Forest Development Plan (OPR 15 (7) (b)). A Forest Development Plan (OPR 15 (2)) and a silviculture prescription (OPR 39 (2) (w)) must contain a reasonable assessment of non-timber resource values known to be on or adjacent to the plan area and must describe the actions to be taken to accommodate those values.

Stand level practices in this chapter are the recommended minimum requirements needed to meet the structural characteristics of natural openings (OPR 21(3)(b)).

When wildlife tree patches (group reserves) are larger than 2 ha (a patch that is isolated within the cutblock boundary and not included within the net area to be restocked) and also meet the age and structural requirements of old seral forest (see "Establishing landscape unit biodiversity objectives") then these larger, within-block patches can contribute to old-seral stage forest requirements within the landscape unit and be used in landscape level retention calculations (OPR 39 (3)(d)). A wildlife tree patch is synonymous with a group reserve in silvicultural terminology.

Given the high degree of ecological variability in our forests, managers need to consider biological diversity on a site-specific basis in order to most effectively apply the recommendations presented in this chapter. This variation may require exceeding the minimum retention requirements recommended in this guidebook.

The recommendations for maintaining biodiversity are an important component of ecosystem health and will have to be integrated with other objectives for forest health.

MAINTAINING STAND STRUCTURE

Stand level recommendations are designed to maintain or restore, in managed stands, important structural attributes such as wildlife trees (including standing dead and dying trees), coarse woody debris, tree species diversity, and understorey vegetation diversity.

SAFE WORK PRACTICES

Safe work practices, as established in conjunction with Workers' Compensation Board, must be followed at all times when implementing the recommendations in this guidebook. Forest workers should have freedom to remove obstacles whenever necessary to maintain a safe working environment. Trees that are marked to leave or are outside the cutblock boundaries can be felled for safety reasons, but should be left on the ground as future coarse woody debris. This will retain the benefit these trees can have on site and remove any incentives for felling and harvesting.

Wildlife trees

A wildlife tree is any standing live or dead tree with special characteristics that provide valuable habitat for conservation or enhancement of wildlife. These trees have characteristics such as large size (diameter and height) for site, condition, age, and decay stage; evidence of use; valuable species types; and relative scarcity. They serve as critical habitat (for denning, shelter, roosting, and foraging) for a wide variety of organisms such as vertebrates, insects, mosses, and lichens.

Maintaining wildlife trees within harvest and silviculture units can be ecologically beneficial in a number of ways. While standing, they provide habitat for many species (birds, bats, and other small mammals) that perform roles in maintaining ecosystem functions. Standing green trees can provide for future wildlife tree recruits.

Wildlife trees will, over time, become sources of coarse woody debris and finally, through decay and nutrient cycling, become incorporated into second-growth forests.

Wildlife tree management strategies can range from the retention of existing wildlife trees, as scattered individuals or in patches, to the creation of new wildlife trees. Many approaches can be applied within a single cutblock, though retention of patches is recommended as the priority approach in most cases. Wildlife tree requirements apply to the use of all silvicultural systems.

Wildlife trees patches (group reserves) and individual live tree retention

Wildlife tree patches (WTP) provide several advantages over other retention strategies. Snags or other potentially dangerous trees are more easily retained in patches than as individual trees and operational inconvenience is minimized. There is also evidence that clumps of trees provide better habitat for birds than do scattered individual trees, as well as an area of relatively undisturbed forest floor within cutblocks. However where scattered individual wildlife trees already exist they should be retained.

Wildlife tree patches should be well distributed across the landscape. The maximum distance between WTPs (500 m) has been based on territory size and dispersal requirements of wildlife.

Recommendations:

Area and distribution of patches or individual trees:

- The retention of wildlife trees should be based on the pre-activity assessment of the wildlife tree values and requirements on or adjacent to the proposed cutblock; and on the described actions that are required to accommodate these values.
- The amount of retention is based on the proportion of each biogeoclimatic subzone in the landscape unit that is operable, and the degree of development that has occurred before the recommendations presented here are applied. The higher

the proportion of operable area in a landscape or the more that previous development has reduced wildlife tree abundance, the greater the amount of retention required.

- Where landscape unit objectives have not been established, the application should be based on an interim landscape unit or a portion of a forest development plan that forms a contiguous geographic unit.
- Suitable areas outside the cutblock, such as riparian management areas, can contribute to the required retention provided the inter-patch spacing requirements are met, and they are mapped and designated as wildlife tree patches in the forest development plan. It is assumed that up to 75% of wildlife tree patch area requirements on the coast, and up to 50% in the interior, will be met in riparian management areas and other constrained areas.
- Good candidates for retention: individual live wildlife trees that provide special wildlife habitat value (such as nest trees); veterans; and other large trees. Patches of deciduous or unmerchantable trees provide opportunities to accomplish this objective. Defective trees of the largest size are often most valuable. These large, live, unhealthy trees (Class 2 trees that are well branched) are important as future habitat since they provide recruitment wildlife trees over time.
- Wildlife tree patches can range in size from individual trees to patches of several hectares. Individual trees, outside patches, can contribute to the required retention area on a basal area equivalency basis. However, simple stem basal area does not equate to patch area. For example, a 1 ha patch containing trees, shrubs, rocky or wet sites, and other ground level diversity is not ecologically equivalent to "x" number of single green stems (such as seed trees) distributed around a harvest block. Therefore, the use of individual trees to contribute to required retention area targets must be planned and selected carefully on a site-specific basis.

- Single green trees chosen for retention should exhibit some characteristics of a valuable recruitment wildlife tree (such as being large and well branched). In order to optimize their habitat value within the harvest block, these trees should be located, wherever possible, near areas that already contain some structural elements of stand level diversity (for example, near a riparian area, rocky outcrop, gully, hardwood patch, or meadow opening). Single live trees left near a shallow gully running across a cutblock are more valuable ecologically than those same green trees distributed randomly throughout the block.
- Wildlife tree patches should be mapped and recorded as part of the documentation of the silvicultural or stand management prescription for the cutblock and be removed from the net area to be restocked.
- No timber harvesting is permitted in designated wildlife tree patches.
- The importance of wildlife tree patches within cutblocks increases with the size of the cutblock. Patches should generally be centered around the most suitable trees and distributed throughout the cutblock, with distances between patches (or other suitable leave areas outside the block) not to exceed 500 m.
- Patches should be located in a way that minimizes windthrow risk. Advantage can also be made of locating patches along small streams, wet areas, gullies or other locations that are likely to pose harvesting, yarding or regeneration problems.
- When partial cutting silvicultural systems are used, sufficient leave trees should be retained throughout the rotation, to meet retention objectives.
- Wildlife trees must be retained at least until other suitable trees can offer equivalent replacement habitat and structural diversity. In most cases this will take at least one rotation.

Patch and live tree retention characteristics:

- A range of tree diameters should be included within patches, favouring larger stems when possible. Care should be taken to include the upper 10% of the diameter distribution of the stand (taken from cruise data), because these are the most valuable wildlife trees.
- Both live and dead trees (subject to safety requirements) should be included in patches representing a range of wildlife tree classes.
- A variety of tree species, including deciduous, should be represented.
- When possible, trees showing wildlife use or presence of heart rot, and those with a large size and well-branched structure should be retained.

MANAGEMENT PRINCIPLES FOR WILDLIFE TREES

- Wildlife tree management includes both the retention of suitable wildlife trees at the time of harvest and during silvicultural activities, and the provision for recruitment of suitable replacement wildlife trees over the rotation period.
- Generally, the most operationally feasible and biologically advantageous method for retaining wildlife trees is to leave patches of live and dead trees as no work zones and to exclude these from the net area to be reforested.
- The amount of wildlife tree area required for a specific cutblock depends on the level and distribution of existing and planned harvesting in the surrounding landscape.

This is a one-time calculation for each biogeoclimatic subzone within the landscape unit (or interim landscape unit or portion of a forest development plan forming a contiguous geographic unit) unless the landscape unit objectives change, a new landscape unit is designated, or operability limits change (changing the area available for harvest). A separate prescription is made for each subzone within the landscape.

X-axis numbers (columns) are the proportion of the subzone within the landscape unit (or interim landscape unit or portion of a forest development plan forming a contiguous geographic unit) that is identified as available for harvest (that is, not inoperable or in some sort of reserve status, such as riparian or protected area).

Y-axis numbers (rows) are the proportion of the available landscape (above) that has already been harvested without application of this guidebook's recommendations or similar prescriptions.

Example: For each biogeoclimatic subzone in the landscape unit, calculate the area available for harvest (the X-axis).

For example, if 30% of the SBSmc area is available for harvest, then, using the 30% column, the recommended minimum proportion of each cutblock to be managed for wildlife trees is between 1 and 9%.

If 50% of the available area has already been harvested without application of these or similar guidelines (Y-axis), then 5% of each new cutblock would need to be left for wildlife tree patches. This can be distributed adjacent to cutblocks in riparian or other long-term leave areas when feasible.

Creating wildlife trees

One method of creating wildlife trees is to high-cut stumps during the use of feller bunchers. This creates standing dead trees called "stubs". These provide structure within second-growth forests and create future coarse woody debris. They cannot, however, replace all attributes associated with full height wildlife trees and thus can not be used as a complete substitute.

Recommendations:

- If mechanical harvesters are being used, snags and cull trees that must be felled should be left as high as can be reached safely. For dead trees, the maximum allowable height must be according to Workers' Compensation Board regulations.

- Stems suitable for stub creation should have some visible defect (such as canker, scar or conk) in the lower bole and little or no lean.
- Creation of wildlife trees by topping, blasting, and other methods should also be considered.

Other stand structure recommendations

Any action that encourages structural diversity within managed stands can have value to biodiversity.

Recommendations:

- Some vertical and structural variability should be maintained throughout stands or rangelands. Within larger stands, this can be achieved by the use of variable stocking densities at planting, in combination with spacing or thinning activities. In rangelands, components of shrub communities should be retained.
- In vegetation management or thinning treatments, patches or strips can be left unthinned on some sites so that dense patches develop.
- Wider spacing of patches can be used to maintain a partially open canopy that will promote understorey vegetation.
- Non-merchantable defect trees should be left as recruitment snags rather than removed.
- Patches of advance regeneration can provide structural diversity.
- Range management should be designed to maintain a component of shrubs within rangelands, based on the shrub component found in undisturbed rangelands.
- Forest encroachment on rangelands can be reduced through management practices such as prescribed fires.
- During tree pest and disease treatments, some areas should be left untreated.
- Openings should be irregularly shaped to most closely reflect natural disturbance patterns.

MAINTAINING TREE AND VEGETATION SPECIES COMPOSITION

The maintenance of the diversity of naturally occurring plant species is key to the maintenance of biological diversity within landscape units. Within cutblocks, several actions can help maintain that diversity.

Recommendations:

- The variety of native understorey plants and plant communities should be maintained across the landscape units.
- Vegetation management treatments can be designed to create variability among or within treatment areas. Effects on non-target plants should be minimized.
- A component of the deciduous species, both immature and mature, should remain after harvesting, spacing, vegetation management, and site preparation activities.
- Extensive conversion from climax to young seral species (such as Douglas-fir to lodgepole pine) or from young seral to climax should be avoided.
- Where suited to the site, stands should be regenerated with a mixture of tree species (natural and planted) rather than with a single species.
- Where mature hardwoods form a minor component of the stand (<20%), greater emphasis should be placed on maintaining these either singly or in clumps.
- Minor tree species such as yew, birch, alder, aspen and cottonwood should be maintained.

MAINTAINING COARSE WOODY DEBRIS

Maintaining coarse woody debris after harvesting is a critical element of managing for biodiversity. However, it is recognized that this requirement conflicts with existing utilization standards. Work is under way to resolve this policy conflict. Until this issue is resolved, utilization standards will take precedence over requirements for coarse woody debris.

Despite this policy conflict, some existing practices can be modified to help address the requirements for coarse woody debris. For example, preliminary indications are that post-logging residue and waste can meet the volume requirements for coarse woody debris if it is well distributed across the entire stand. This will not be the case in situations of whole-tree harvesting, clean site preparation practices, or excessive salvage of material not considered merchantable under current utilization standards.

Recommendations:

- Modify whole-tree harvesting by limbing and topping on site.
- Maintain residue and waste well-distributed across the stand (avoid practices such as piling and burning).
- Leave non-merchantable material on site.

3

Biodiversity of Living Organisms

INTRODUCTION

The diversity of living organisms we observe today is the product of billions of years of evolution. This biodiversity is, however, now threatened by humanity. Global trends show that human activities are destroying and degrading a diverse range of ecosystems and result in the extinction of thousands of species annually. Wilson (1992), among other leading commentators, calls this a great spasm of extinction - caused entirely by humans.

Australia has a diverse and often unique environment that represents a priceless heritage that should be a source of pride to all Australians. The Australian government is a signatory to the United Nations CBD, and Australia has a national strategy for safeguarding its biodiversity heritage. Some aspects of the Australian environment were in relatively good condition by international standards (SoE 1996) and the approach to environmental management had international recognition in some areas.

The 1996 Report also demonstrated that Australia has some very serious environmental problems, the cumulative consequences of human population growth and distribution, lifestyles, technologies and demands on natural resources over the last 200 years and more. The Report suggested that changes were needed in government policies and programs, corporate practices and personal behaviour. The estimated population of Australia when Europeans arrived varies greatly, from 300 000 to 1.5 million. Seventy years later, the European population had reached one million. By 2001, the total population approaches 20 million. These citizens and their governments, industry and community organisations have responsibility for Australia's biodiversity.

STATE OF THE ENVIRONMENT REPORTING AND THE 2001 NATIONAL REPORT

Progress towards sustainability is difficult, if not impossible, without adequate and accessible information about the environment. State of the Environment (SoE) reporting can be a powerful tool for providing this information - to the public, industry, non-government organisations (NGOs) and all levels of government. As such, SoE reporting is being embraced at the local, regional, state and national levels. It allows regular reports on agreed sets of indicators of changes and trends in environmental conditions, in much the same way as well-accepted economic indicators are used to report on the state of the economy. It describes the effects of human activities on the environment, and their implications for human health and economic wellbeing. It also provides an opportunity to monitor the performance of government policies against actual outcomes. Thus, SoE reporting can act as a report card on the condition of the environment and natural resource stocks.

THE MEANING, SIGNIFICANCE AND IMPLICATIONS OF BIODIVERSITY

What is Biodiversity?

After billions of years of evolution, earth is home to a large array of life forms and ecosystems. The Australian continent supports a significant proportion of this global diversity, and Australians increasingly accept a share of the responsibility for it. Concern for non-human life forms goes back in history and across many cultures. However, the level and nature of concern for these life forms changes over time, reflecting our understanding of the nature and importance of this legacy, the values we ascribe to it, the threats it is under and what might be done to conserve it (see Wilson (1992) and Burgman & Lindenmayer (1998)).

Defining biodiversity

The evolutionary legacy of life on earth is now described by the term 'biodiversity', or biodiversity, but this term has come into wide use only recently. It refers to the variety of life of earth - plants, animals and microorganisms, as well as the variety of genetic material they contain and of the ecological systems in which they occur. It is a simple concept, but one which also has great complexity and significance.

In Article 2 of the 1992 United Nations CBD (United Nations CBD 1992a), biodiversity was defined as:

> *the variability among living organisms from all sources including, inter alia, terrestrial, marine and other aquatic ecosystems and the ecological complexes of which they are a part; this includes diversity within species, between species and of ecosystems.*

This definition was repeated in Australia's *Environment Protection and Biodiversity Conservation Act 1999* (EPBC Act), and the NSCABD (Commonwealth of Australia 1996) expands on the three levels at which biodiversity occurs. These levels of biodiversity are:

1. genetic: the variety of genetic information contained in all of the individual plants, animals and microorganisms that inhabit the earth - genetic diversity occurs within and between the populations of organisms that comprise individual species as well as among species
2. species: the variety of species on earth
3. ecosystem: the variety of habitats, biotic communities and ecological processes.

These three levels of diversity are interrelated and interdependent (e.g. a population of a species is thoroughly dependent on its habitat for survival, and a functioning ecosystem is dependent on the complex of species that comprises it).

ATTITUDES TOWARDS BIODIVERSITY

In the human history of Australia, changing values, scientific knowledge and cultural understanding have altered the way we perceive and interact with the natural environment generally and with biodiversity in particular. The following six phases offer a simplified summary of this process, characterising six different sets of attitudes (extended from Frawley 1994; see also Griffiths 1996; Bowman 1998). While the emphasis has changed over time, these phases reflect attitudes toward the Australian environment and biodiversity that are, to some extent, current.

Indigenous

The Indigenous peoples of Australia have valued and utilised components of biodiversity for at least 60 000 years (Thorne et al. 1999). Indigenous culture and practices have developed in a dynamic relationship with the environment. For example, the use of fire and hunting of animals helped shape the terrestrial environment. Today, Australians live in a cultural landscape that incorporates a diversity of manifestations of the interactions between humans and the natural environment over the last 60 000 years.

While 'western science' is still dominant in terms of the way non-Indigenous people view, describe and classify our flora and fauna, there are an increasing numbers of examples where Indigenous ecological knowledge is being accepted on an equal basis (Kakadu Board of Management and Parks Australia 1998).

Exploration and contact

Australia's flora and fauna were often a focus, but mostly in terms of their possible value for colonial trade or of their scientific interest and peculiarity.

Exploitative pioneering

The environment was viewed as feedstock for colonial economic and trade development. There was little in the way of specific law or management aimed at the protection of native species.

Acclimatisation Societies were set up so the early settlers could make the environment more like 'home' and brought in some plants and animals that ended up becoming pests.

Wise use for national development

New resource management arrangements were established that were informed by science and aimed at management of natural resources (especially forests, water and soils) to best answer human needs in both the present and future.

Modern environmentalism

The modern conservation movement, prominent from the 1960s onward, expressed new issues and values in environmental management debates. In this era, the intrinsic value of biodiversity was more widely recognised, and when laws and policies for biodiversity conservation became the norm rather than the exception.

Ecologically sustainable development

Since the early 1990s, the central organising concept used by governments to describe human interaction with the environment has been 'sustainable development', termed 'ecologically sustainable development' (ESD) in Australia. The main aims of ESD are to: integrate environmental, social and economic concerns over a long time; adopt a precautionary approach; and recognise the importance of biodiversity and ecosystem processes. Current debates and practices still feature these different attitudes. One constant is the character of Australia's biota. The biota, special to Indigenous people for millennia, appeared fantastic and fascinating to early European observers. Beneath the superficial strangeness of kangaroos, black swans, platypus and endless eucalypts, a special quality has been increasingly realised.

DOMESTIC LAW AND BIODIVERSITY

One indication of change is the expression of biodiversity in domestic law. Integrated nature conservation legislation, which caters for harvesting control, conservation reserves and species protection, and specific legislation, which deals with threatened species, only date back a few decades in Australia (most states enacted such laws in the 1970s). By 1999, over 120 state, territory and Commonwealth statutes expressly referred to ESD as an objective and set of guiding principles (Stein 2000). There has also been a growth in the number and scope of international instruments concerning biodiversity.

VALUES ASSOCIATED WITH BIODIVERSITY

The following categories summarise the different values people and society place on biodiversity:

Direct utilitarian value

Biodiversity is consumed by humans as food and is used to feed stock. It provides materials such as timber and fibre, medicines, chemicals and genetic material.

Indirect utilitarian values

Indirect utilitarian values include the maintenance of 'ecosystem services' or important ecological processes. Examples include maintaining water quality in catchments, moderating atmospheric processes or weather, conserving the structure or fertility of soil, maintaining coastal function, assimilating or removing wastes from water or soil, maintaining evolutionary potential in ecosystems, sequestering carbon emissions, cycling of nutrients, pest control, and pollination of crops.

Aesthetic and recreational values

Biodiversity has aesthetic and recreational uses for humans, both in the form of specific taxa such as flowers, birds, trees or whales, and as components of natural or semi-natural landscapes such as the Great Barrier Reef and the wetlands of Kakadu National Park.

Scientific and educational values

Scientific discovery can lead to the development of utilitarian values. It will often be through scientific research, other forms of investigation and learning about community or Indigenous knowledge that such uses will be recognised. Also, the variety of life is of educational value across a wide variety of subjects and disciplines (e.g. biology, biochemistry, ecology, genetics and agronomy).

Intrinsic, spiritual and ethical values

Various cultural and religious systems (e.g. Aboriginal and Torres Strait Islander people) place value on components of biodiversity. Also, there is the ethical position that non-human forms of life have intrinsic value and a right to exist independent of any use to humans.

Future or 'option' values

For all of the above values, there is the added dimension of keeping options open for the future. We are uncertain as to what species and populations are crucial to ecosystem services, or the actual significance of some of these services. Similarly, there may be uses for species or genetic diversity yet to be discovered, such as for food or medicine. And, if the values held in society change as they have in the past, then what is viewed as unimportant now may be more highly valued in the future.

All the values identified above are evident in Australian society, and many individuals will value biodiversity for more than one of these reasons. Perhaps the most important change in understanding in the long term has been the recognition of the reliance of biodiversity on functioning ecosystems, and its role in maintaining ecological processes. This recasts biodiversity science, policy and management in important ways. Managing just a few species and protecting a small selection of natural areas is not sufficient to protect Australia's biodiversity.

Another major and continuing change is the attention being paid to *indirect* (or underlying) as well as *direct* (or proximate) causes of biodiversity loss. For example, land clearing by farmers is a direct cause of biodiversity loss in Australia. The indirect causes lie in the social, institutional and economic settings that influence farmer behaviour and farm profitability. This includes the information available to landholders, economic conditions affecting rural industries and perverse incentives encouraging clearance.

This shift in emphasis deepens our understanding of the processes of biodiversity loss and allows more sophisticated policy responses. In the land clearance example, strict regulation is invited by the direct cause, whereas understanding the indirect cause invites the use of incentive mechanisms, forward planning, information provision and other approaches.

CONCERN FOR BIODIVERSITY

An indication of society's concern for biodiversity can be gleaned from opinion surveys, even though these have only been undertaken recently. An analysis of national surveys from 1975 to 1994 identified high levels of concern over environmental issues, and biodiversity was consistently important within the broader environmental field (Lothian 1994). During the 1990s, the ABS undertook more regular surveys (ABS 1999a). In 1999, the environment was nominated as the most important social issue by 9% of the Bureau's sample, above issues such as crime and health. Of those who did not rank the environment as their top issue, 69% stated that they were concerned about environmental issues. This represents a slight decrease on recent years although in 1986 the corresponding figure was 49% (SOE 1996, p. 10-11). The issues of concern most directly relevant to biodiversity, destruction of trees/ecosystems and of animals/wildlife, were identified by 29% of people, and 43% believed that the quality of the environment had decreased in the 1990s.

AUSTRALIA'S BIODIVERSITY: A UNIQUE HERITAGE

Australia's biodiversity has been especially shaped by the features of the Australian landscape, and long periods of evolutionary isolation. Compared with other parts of the world and because of the age and deep weathering of the landscape, Australia has low relief, low and variable rainfall and low nutrient soils. Long periods of isolation from other landmasses have resulted in the evolutionary radiation of groups of terrestrial plants and animals such as eucalypts, wattle, hummock grasses and marsupials (Strahan 1983; Flora of Australia 1999). The aquatic environment also hosts a great diversity of plant and animal species, although this element of the Australian flora and fauna often receives less attention despite its ecological significance (Flora of Australia 1999).

Geological History

All the present continents coalesced into a single supercontinent, Pangaea, about 230 million years ago. Within Pangaea, the land masses of Australia, New Guinea, South America, Africa, Madagascar, Antarctica, India, New Zealand, Arabia and parts of South-East Asia were close together and formed the continent of Gondwana. Pangaea began to break up about 160 million years ago. Australia and Antarctica parted only 35 million years ago. At that time, the climate of Gondwana was humid and temperate and rainforest covered much of the landscape. Australia drifted north and collided with the islands in the Pacific about 15 million years ago. During this period of isolation, the continent was exposed to dramatic climatic changes. The circum-Antarctic current formed when Australia and Antarctica separated, resulting in substantially reduced temperatures at the southern pole and the formation of the Antarctic ice cap over the next 20 million years. Most of the Australian landscape was stable during the isolation of the last 30 million years, resulting in today's ancient, weathered surfaces (Archer et al. 1998).

These geological events and the associated landscape and climate changes have created a unique legacy of flora, fauna and landforms on the Australian continent. In general, Australia's flora and fauna are a mixture of the original Gondwanan stock, modified over millions of years of separation, and the more recent arrivals from Asia. For example, in northern Australia, several plant and animal groups are shared with the islands of New Guinea and Indonesia. Many taxa (families and genera) are also shared between southern Africa, South America and Australia because of their common origins. For example, marsupials are common in Australia, South

America and New Guinea, but not in Asia. The rainforest trees *Nothofagus* and *Araucaria* occur in South America and Australia. The flora of south-west Australia shares many relatives with the flora of southern Africa.

DROUGHTS AND FLOODING RAINS

The major components of the Australian terrestrial flora and fauna have diversified into a rich spectrum of species adapted to the vagaries of the Australian environment. The droughts, fires and floods that are a part of the Australian environment are closely tied to regional climate patterns, especially the El Nio-Southern Oscillation (ENSO) phenomenon. Rainfall, especially for eastern Australia, has a very high interannual variability. Extensive and often prolonged droughts, sometimes with severe fires in forested regions, occur during El Nio years, whereas flooding is common in La Nia years. The widespread rainfall associated with La Nia events affects the abundance of plants and animals through the filling of water bodies such as Lake Eyre and indirectly through its effect on fire regimes and pest outbreaks. Disturbance and climatic cycles are also important for biodiversity in marine systems. Natural agents of change in marine communities include tropical cyclones and temperate severe storms, outbreaks of predators and ENSO effects on currents and coastal ocean productivity.

Megadiverse Countries

Australia is identified as a megadiverse country. The concept of megadiversity is based on the total number of species in a country and the degree of endemism at the species level and at higher taxonomic levels. The World Conservation Monitoring Centre recognised 17 megadiverse countries in July 2000 including Australia, Brazil, China, Colombia, Democratic Republic of the Congo (DRC) (formerly Zaire), Ecuador, India, Indonesia, Madagascar, Malaysia, Mexico, Papua New Guinea, Peru, the Philippines, South Africa, the United States of America (USA) and Venezuela. Together, these 17 countries harbour more than 70% of the earth's species.

Figure 1: Number of endemic vascular plant species in 17 megadiverse countries illustrating that Australia has the fifth highest number of species in this group.

PNG, Papua New Guinea; DRC, Democratic Republic of the Congo.

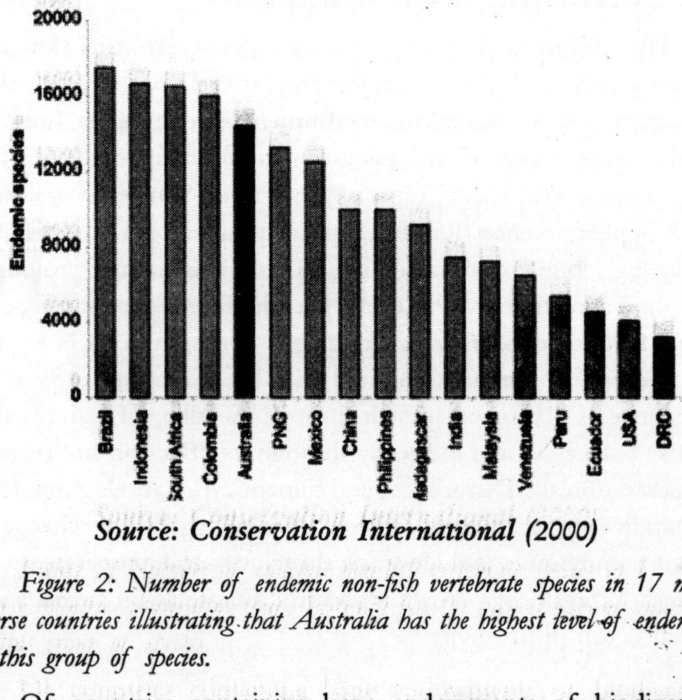

Source: Conservation International (2000)

Figure 2: Number of endemic non-fish vertebrate species in 17 mega-diverse countries illustrating that Australia has the highest level of endemism for this group of species.

Of countries containing large endowments of biodiversity, Australia is unique in another, very significant way. Of all the countries classified as megadiverse, Australia is one of only two countries in the high income category.

Australia as a richer, healthier, better educated country with a more developed economy is better positioned to deal with problems of environment and development than all but one other megadiverse nation - the USA. This position carries a special responsibility and implies that a high standard of biodiversity protection can be expected in Australia. It also carries with it an opportunity for world leadership.

PNG, Papua New Guinea; DRC, Democratic Republic of the Congo.

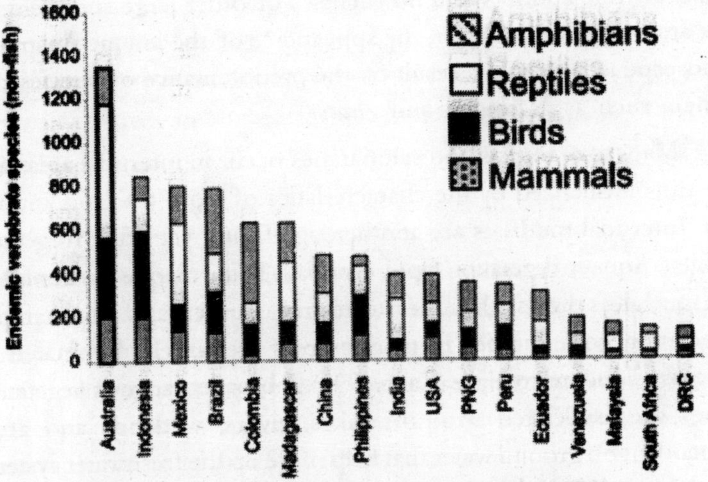

Source: Conservation International (2000)

KEY FEATURES OF AUSTRALIA'S BIODIVERSITY

Vegetation Types

The terrestrial ecosystems found in Australia are very varied owing to the range of climates - including subalpine, cool temperate, arid and tropical biomes. This diversity leads to a range of vegetation types whose distribution at a broad scale is limited mainly by rainfall. These include the spinifex-dominated arid interior, semi-arid shrublands, tropical and temperate grasslands, rainforests and woodlands (including savannas), eucalypt-dominated forests and shrublands, chenopod shrublands, heathlands, alpine and subalpine vegetation (Groves 1994; Flora of Australia 1999).

The only available Australia-wide map of Australia's natural vegetation was compiled by JA Carnahan (Commonwealth of Australia 1990). During 2001, a new continental-scale vegetation map is to be produced by the National Land and Water Resources Audit (NLWRA). This is part of a major initiative, the National Vegetation Information System, which is a collaborative project between the state, territory and Commonwealth governments. Australian terrestrial vegetation contains very few deciduous species, relatively few conifers and no cactuses or other large succulents in the arid zone. In particular, the appearance of the unique Australian landscape is largely as a result of the predominance of species-rich genera such as *Eucalyptus* and *Acacia*.

Mangrove forests and saltmarshes occur in intertidal areas and are thus influenced by the characteristics of both the land and the sea. Intertidal mudflats are another significant, yet often forgotten, habitat. Aquatic vegetation types, especially in the marine environment, are much less diverse than their terrestrial counterparts. For example, vegetation communities in marine areas are largely dominated by seagrasses and macroalgae (Zann 1995). A broader range of vegetation types are associated with the lakes, rivers, wetlands and areas dependent on ground water that help make up the freshwater systems of Australia (Boulton & Brock 1999).

The photographs below illustrate an indicative range of diversity of vegetation types. Some of these vegetation types such as tropical rainforests and alpine vegetation are restricted in occurrence. This contrasts with tropical savannas (dominated by eucalypt species) and spinifex-dominated grasslands which cover large areas of tropical and arid Australia, respectively. Another widespread community is the Mulga (*Acacia aneura*)-dominated shrublands. Together with hummock grasslands wooded with Mulga, these communities are estimated to occupy 1.5 million square kilometres or about 20% of the Australian continent (Hodgkinson 2001). Although most of the continent is dominated by semi-arid and arid ecosystems, ecosystem diversity is greatest in the higher rainfall regions on the eastern and southern edges of the continent.

Marine Habitats

Australia has one of the world's longest national coastlines and one of the largest marine jurisdictions, about twice the size of the Australian mainland ranging from the sub-Antarctic to the tropics.

These measurements refer only to surface area - marine organisms live in all available habitats extending throughout the water column to the floor of deep ocean trenches. Some of these trenches reach 11 000 m in depth but even in these extreme habitats, marine invertebrates still occur. In addition to well-known habitats such as shorelines, estuaries, rocky shores and rock pools, coral reefs and seagrass beds, marine invertebrates can be found in other habitats such as the interstitial spaces between grains of sand, around hydrothermal vents, on floating debris (e.g. algal mats and driftwood attached to other animals), the peaks and slopes of seamounts and swimming or floating in the water itself.

Classifying Vegetation or Habitat Types

Indigenous people have developed classifications for vegetation or habitat types that can bear marked similarities to the broad associations defined by western science. While western science is based on evolutionary theory and Linnaean taxonomy, all aspects of Indigenous life is governed by the genesis of life and classification based on their religious and social laws (Baker & Mutitjulu Community 1992).

Table: Indigenous words for arid zone systems and taxa

Habitat type	*Anangu name*
Rocky range and outcrops	puli
Mulga shrublands	puti
Riverbed and riverbanks	karu
Sand plain	pila
Dunefields	tali
Saltlakes or claypans	Pantu or tjintjira

THE BIODIVERSITY CHALLENGE: RESPONSIBILITIES ROLES AND PARTNERSHIPS

In seeking to reconcile economic political and legal systems with the natural systems on which they depend, several core 'attributes' are encountered and these very much shape the way in which policy management and scientific responses might be framed.

First, ESD issues need to be considered over different space and time scales from those of many other policy issues. Ecosystems and ecological processes rarely concur with political boundaries, national borders, election cycles or the timing of government budgets. Second, there are significant knowledge gaps concerning biodiversity: its component parts, the role of taxa in ecosystems, and both the actual and possible future values of biodiversity. Third, ESD problems often involve cumulative effects (e.g. diffuse pollution sources affecting an ecosystem over time, and dryland salinity) and irreversible effects (e.g. species extinction). Fourth, many environmental issues cut across industry and government portfolio sectors, which means that responses must be integrated and coordinated. Finally, there is the sheer novelty of many ESD issues - their relatively recent arrival on the policy agenda means that there is a lack of defined policy and property rights and responsibilities and a lack of agreed management approaches and policy instruments. These attributes characterise the challenge of biodiversity conservation.

IMPROVING INFORMATION AND KNOWLEDGE

Several elements are involved in trying to improve the information and knowledge base on the state of Australia's biodiversity. These include the tasks of improving taxonomic knowledge of Australia's biodiversity; improving our ecological knowledge across populations, species and ecosystems; describing the abundance, status, life histories of these and the threats to them; and the more recently defined area of increasing our understanding of genetic diversity within species and populations. All these require collaboration between scientific and government institutions as well as the important role of the private sector in research and development.

If our understanding of biodiversity is to inform policy and management, then an historical dimension is needed to provide the baselines against which judgments can be made which brings in a wider range of research disciplines (i.e. history) as well as members of the community and Indigenous groups who have strong links with specific places.

Information regarding the success or otherwise of past and present policy interventions and management practices is important if we are to know whether and how well these are working. Although governments are central in monitoring policy and management, the community, science, relevant professions and industry also need to be involved.

COMMUNICATION AND EDUCATION STRATEGIES

Gathering and storing information on the protection and sustainable use of Australia's biodiversity is pointless without its communication to those who need to use it. This includes the spread of research findings to decision makers, other researchers and the broader community. The media, schools, university and various training curriculums are also important.

Managing Biodiversity

Managing biodiversity across the Australian landscape requires:

- management of protected areas such as national parks and nature reserves
- management of biodiversity outside the reserve system
- control of harvesting and use of biota.

Although reservation of land is a government responsibility, many reserves are established following community pressure. Community involvement in the management of reserves has increased through boards and committees that include community representatives, as in the case of those co-managed by Indigenous land owners and park services (e.g. Kakadu Board of Management and Parks Australia 1998; Uluru-Kata Tjuta Board of Management and Parks Australia 2000). The off-reserve task involves a mix of legislation, regulation, incentive measures and voluntary efforts across farmland, in production forests, in industries such as tourism and mining, on Indigenous lands and in the marine realm. The control of harvesting and other uses of native plants and animals applies to both public and private land and waters within a regulatory framework that involves close interaction with individuals, communities, firms, governments and industry.

Restoration and Recovery

An important aspect of managing biodiversity is the increasing attempts to restore habitats to some previous or more desirable condition and to encourage the recovery of diminished or threatened species or populations. As well as government agency staff and scientists, community groups, private landholders and firms have an important role in many such attempts. While *in situ* conservation is the preferred option, in some instances it is not always feasible.

POLICY AND MANAGEMENT SETTINGS

Protection of biodiversity is primarily the responsibility of governments. The broader community, however, has input in various ways into policy formulation, the design or reform of laws and on-the-ground implementation of policies.

In the case of biodiversity and most other environmental issues, legislation (i.e. statute law) rather than common law has overwhelmingly been the predominant source of power and responsibility. Legislation does or can serve many purposes, such as: setting out government objectives and the structure of agencies; establishing opportunities for public participation; enabling financial incentives; defining procedures for development approval; impact assessment; strategic planning processes; and ensuring transparency of decision making.

Australia's federal system of government is important to understanding how biodiversity is or can be managed. Under Australia's constitutional arrangements, the great bulk of responsibility for environmental management, including biodiversity, resides with the states. The Commonwealth, however, has considerable potential and actual power via the trade, external affairs, corporations, finance and taxation and race powers in s51 of the Australian Constitution (Bates 1995). These powers were confirmed in court cases over the 1970s and 1980s. Over the past decade, however, there has been a trend toward less confrontational and more cooperative approaches to the environment. The Intergovernmental Agreement on the Environment, the National Environment Protection Council and the development of national policies such as the NSCABD reflect this.

A major role of the Commonwealth, recognised in the Intergovernmental Agreement on the Environment, is in the implementation of international agreements (e.g. the Convention on Biodiversity, CBD). Some international obligations have become triggers in the EPBC Act, which can lead to environmental assessment and regulation of project proposals by the Commonwealth. The triggers of relevance to biodiversity are significant effects on Ramsar wetlands, World Heritage areas, migratory species, nationally threatened species and ecological communities and Commonwealth marine areas. In early 2001, the Environment Minister, following advice from the Threatened Species Scientific Committee, listed land clearing as a key threatening process to biodiversity.

LAW IN BIODIVERSITY CONSERVATION

The following examples demonstrate the increasing relevance and role of law in biodiversity conservation:

- very different approaches taken in the different States to biodiversity management and the tensions which this causes (e.g. Queensland's vegetation clearance legislation proposes to offer compensation for land use restrictions but this has been strenuously resisted in New South Wales)
- movement away from ad hoc regulation of individual project proposals towards strategic planning (e.g. bioregional planning)
- law's role in setting up decision-making procedures relating to biodiversity management rather than setting out absolute

standards (which would guarantee the protection of threatened species)

- lack of integration between older law focusing on the management of particular resources (e.g. water) and more recent law based on a holistic view of the environment
- lack of integration between the functions of elected local councils and appointed community resource management committees (e.g. catchment committees)
- perceived failure of command and control regulation specifically in the private land context and an attempt to develop economic instruments and community driven mechanisms (e.g. catchment committees)
- increased recognition that biodiversity conservation is as much about active management as land use restriction
- difficulties posed by planning law's traditional respect for existing land uses.

BIODIVERSITY STATUS, TRENDS AND INDICATORS

The 1996 State of the Environment Report

- The 1996 Report (SoE 1996) was motivated by a commitment to ecological sustainability and a concern for Australia's biodiversity. It outlined over 50 major issues of concern for biodiversity and its conservation including 11 key issues that were identified as threats to biodiversity.
- The threats to biodiversity identified in SoE (1996) are presented here in further detail. Many ecosystems and species are threatened by human activities such as habitat clearance and modification and the invasion of systems by exotic organisms. Genetic diversity is also threatened although the nature and detail of genetic diversity loss is most poorly documented. At the general level, SoE (1996) stated that 'to balance conservation of biodiversity, human population growth and economic development' would require 'substantial changes in the way that land and oceans are managed'.

Clearly between 1996 and 2001, an expectation of 'substantial' change would be likely to be disappointed. Mostly we do not yet fully understand what such changes entail. This Report shows that there have been both encouraging signs of improvement and evidence of both lack of progress or emerging and as yet poorly addressed issues.

INDICATORS OF BIODIVERSITY

Environmental Indicators

This Report pioneers the use of environmental indicators on a continental scale. Environmental indicators are physical, chemical, biological or socioeconomic measures that are considered to best represent the key elements of a complex ecosystem or environmental issue. When fully developed they should help define the nature and size of environmental issues, set goals for their solution and track progress towards these goals (Heinemann et al. 1998). In order to track these changes, a monitoring program is essential where repeated sets of measurements are compared with a benchmark set or condition. For SoE reporting, these benchmarks must enable the effects of current programs and policies and of land/resource management activities to be assessed in relation to their biodiversity outcomes.

The 2001 State of the Environment Report

The national level biodiversity indicators that form the basis of the report were developed by Saunders et al. (1998). A total of 65 indicators were recommended, 14 of which related to pressures on biodiversity, 17 to the condition of biodiversity and 34 to responses to the loss of or perceived threats to biodiversity.

When Saunders et al. (1998) were developing the national level indicators, each one was assessed to see if it would:

1. serve as a robust indicator of environmental change
2. reflect a fundamental or highly valued aspect of the environment

3. be either national in scope or applicable to regional environmental issues of national significance
4. provide an early warning of potential problems
5. be capable of being monitored to provide statistically verifiable and reproducible data that show trends over time and preferably apply to a broad range of environmental regions
6. be scientifically credible
7. be easy to understand
8. be monitored regularly with relative ease
9. be cost-effective
10. have relevance to policy and management needs
11. contribute to monitoring of progress towards implementing commitments in nationally significant environmental policies
12. where possible and appropriate facilitate community involvement
13. contribute to the fulfilment of reporting obligations under international agreements
14. where possible and appropriate use existing commercial and managerial indicators
15. where possible and appropriate be consistent and comparable with other countries and state and territory indicators.

Reporting Scale

The bioregional scale provides a national framework for the conservation and the protection of biodiversity and is used as the basis for reporting for many of the indicators developed by Saunders et al. (1998). The two major regionalisations used in Australia are the: IBRA (Thackway & Cresswell 1995) and IMCRA (IMCRA Technical Group 1998). In Australia, 85 IBRA and 60 IMCRA regions have been identified. The National Reserve System Program (NRSP) has funded additional studies for the refinement of IBRA regions and in October 2000 a set of revised boundaries was agreed by the Commonwealth, state and territory governments. As a result, IBRA version 5 is being developed and where indicated, is used in this Report.

OTHER NATIONAL SCALE INDICATORS

Two other sets of national scale indicators have or are being developed in Australia. First, ANZECC (2000a) has chosen 13 core biodiversity indicators on the basis that they can be used to report on the state of the environment across jurisdictions within Australia. Many of these indicators have strong links to those recommended by Saunders et al. (1998).

The second set of national scale indicators is being developed by the ABS. These represent a small set of headline indicators based on the core objectives of the National Strategy for Ecological Sustainable Development. Six indicators have been identified under the heading in the ABS document *Protecting biodiversity and maintaining ecological processes and life support systems.* Those most directly relevant to biodiversity conservation focus on the extent and condition of representative ecosystems, the health of land and water systems and the number of extinct endangered and vulnerable species and ecological communities.

INSTITUTIONS AND POLICIES

Changing Roles and Responsibilities

In line with changing understanding of biodiversity and the ways in which it needs to be protected, roles and responsibilities in biodiversity management have changed significantly in Australia in recent years, and continue to evolve. Consideration of all classes of biodiversity indicators - pressure, state, response and implications - are influenced by these changes in terms of who has policy and law-making power, who gathers information and who is involved in management in the field.

The changing roles and responsibilities of the Commonwealth, state and territory governments were outlined earlier. Change has occurred within jurisdictions as well. Traditionally a statutory authority (a parks and wildlife authority or service) was responsible in each jurisdiction for almost all practical wildlife protection and nature conservation activities perhaps with a related policy-making branch in a government department. Some jurisdictions have moved away from this with reserve management becoming an activity carried out by a government department (e.g. the Commonwealth after the EPBC Act, although the statutory office of Director of Parks is maintained).

Another model is where the parks service is incorporated into a broader arrangement. This approach is being used in Queensland where the reserve management agency is part of a larger Environmental Protection Authority (EPA). The responsibility for biodiversity has changed often as well, ranging from an environment department to various combinations of agriculture, forestry, water pollution control, cultural heritage planning and land administration. Increasingly, other portfolios (e.g. water, agriculture) have to some extent incorporated biodiversity into their decisions (whether adequately or not).

These changes in the organisation of the public sector are reflected and in many ways are products of broader changes outside of government. The expansion of focus to include off-reserve conservation matches the move away from single government agencies managing reserves. The increasing inclusion of private landholders and community groups in policy formulation management and program delivery has been a feature of the 1990s, as well as the emphasis on self-regulatory approaches by firms and industry sectors and greater focus on the role of local government.

Partnerships

Not only have the statutory sources of responsibility and the number of organisations protecting biodiversity changed, but also the relationships between these bodies has changed. Few of the initiatives covered in this Report are undertaken by one interest alone. Partnership arrangements are becoming normal rather than the exception and they exist in all combinations between the following:

- governments: state or territory, Commonwealth, local
- community groups representing conservation, development, public health and other interests
- private sector individuals and organisations: landholders, firms, industry associations
- research and scientific practitioners and institutions.

STATUTORY POLICY AND ORGANISATIONAL SETTINGS

Biodiversity conservation in Australia is influenced by an immense array of statutory and policy settings. As it is a deeply cross-sectoral issue, it is doubtful that biodiversity can ever be effectively managed through a small and narrowly focused set of laws and policies.

Only a small proportion of relevant policy and law is subject to the type of monitoring that can provide an accurate evaluation of the effectiveness of biodiversity law and policy. The difficulties of reporting against many 'Response' indicators indicate this (e.g. the number of management plans and research programs and the amount of funding spent on control of exotic species).

MAJOR BIODIVERSITY POLICY, LAW AND PROGRAMS

There is a range of policy and law of direct relevance to biodiversity. Originally regulation of use and trade of wildlife and declaration and management of a reserve estate was the predominant area of activity. While such regulation remains important in recent years, there has been an expansion of legislative and policy activity especially covering threatened species and ecosystems, and management of biodiversity occurring outside of the conservation reserve estate. This broadening reflects changing understanding of biodiversity issues and the increasing range of recognised social values ascribed to biodiversity.

The Environment Protection and Biodiversity Conservation Act 1999

At the Commonwealth level, the *Environment Protection and Biodiversity Conservation Act 1999* (EPBC Act), which came into force in mid-2000, represents a major consolidation of legislation dealing with biodiversity. The *Environment Protection and Biodiversity Conservation Act 1999* box below summarises the major features of the EPBC Act that concern biodiversity. With any law of this kind, and especially in this case, the proof of legislative effectiveness lies in the implementation and this cannot be assessed so early in the life of the EPBC Act. Overall, the EPBC Act is considered by many to be a welcome consolidation of biodiversity law. Critical commentaries on the EPBC Act by industry and conservation interests illustrate some key issues concerning the legislation that may determine the effectiveness of the Act.

Any particular provision of the EPBC Act will be viewed quite differently by different interest groups. Other matters under the Act of relevance to future evaluation and reporting are: the functioning of the key advisory bodies (BDAC, Threatened Species Scientific Committee, Indigenous Advisory Committee), the treatment of biodiversity issues under impact assessment procedures (IAP) whether they are Commonwealth or accredited state or territory ones, and the listing and production of plans for threatened species communities and threatening processes.

The Environment Protection and Biodiversity Conservation Act 1999

The new Commonwealth legislation, the *Environment Protection and Biodiversity Conservation Act 1999* (EPBC Act), came into effect in July 2000. It aims to define, consolidate and streamline the role of the Commonwealth in environmental protection, assessment of environmental effects on nationally significant places and issues, and in biodiversity conservation. The EPBC Act replaces the following existing legislation: *Environment Protection (Impact of Proposals) Act*

1974, National Parks and Wildlife Conservation Act 1975, Whale Protection Act 1980, World Heritage Properties Conservation Act 1983 and the *Endangered Species Protection Act 1992.*

The Objects of the EPBC Act defined at s3 are to:

- provide for the protection of the environment especially those aspects of the environment that are matters of national significance
- promote ESD through the conservation and ecologically sustainable use of natural resources
- promote the conservation of biodiversity
- promote a cooperative approach to the protection and management of the environment involving governments, the community, landholders and Indigenous peoples
- assist in the cooperative implementation of Australia's international environmental responsibilities
- recognise the role of Indigenous peoples in the conservation and ecologically sustainable use of Australia's biodiversity
- promote the use of Indigenous peoples' knowledge of biodiversity with the involvement of, and in cooperation with, the owners of that knowledge.

The EPBC Act primarily deals with actions of the Commonwealth, Commonwealth areas and matters of national environmental significance. Matters of national significance have been initially defined as: World Heritage areas, Ramsar wetlands, nationally listed threatened species and communities, nationally listed migratory species, nuclear actions and the marine environment.

The principal provisions and procedures under the EPBC Act concerning biodiversity conservation are as follows:

- lists threatened species and ecological communities, migratory species and threatening processes, and the preparation of recovery plans, wildlife conservation plans and threat abatement plans (TAPs)

- creates offences and penalties for activities such as killing, injuring and trading of listed species or communities in a Commonwealth area
- establishes the Australian Whale Sanctuary
- allows for regulations to be developed dealing with access to biological resources in Commonwealth areas
- allows for regulations to be developed to list non-native species that do, or may, threaten biodiversity, control trade in those species and make plans for their management or elimination
- provides for conservation agreements between the Commonwealth and landholders relating to biodiversity, including possible assistance for management; these agreements are binding to successor landholders
- establishes procedures regarding the nomination and management of World Heritage areas, Ramsar wetlands and biosphere reserves
- allows for the declaration and management of Commonwealth reserves, including regulation of activities, management plans and public consultation; 'conservation zones' may be declared as an interim measure prior to declaration of a Commonwealth reserve
- establishes the Indigenous Advisory Committee, Threatened Species Scientific Committee and BDAC.

Any project that will have, or is likely to have, a significant effect on a matter of national environmental significance needs to be approved and assessed under the EPBC Act. To end the duplication of assessing projects under both state and Commonwealth laws, bilateral agreements are being negotiated so that projects that trigger the EPBC Act will be assessed under the laws of the relevant jurisdiction. The accreditation process is meant to ensure that national benchmarks are met. The first bilateral agreement under the EPBC Act was signed by the Commonwealth and Tasmania on 15 December 2000 and negotiations with other states are well advanced.

Several amendments have been proposed to the EPBC Act, including adding a greenhouse trigger (the seventh trigger in the Act) as well as amending the world heritage provisions. In early 2001, the Minister for the Environment, following advice from the Threatened Species Scientific Committee, listed land clearing as a key threatening process to biodiversity. Several further amendments have been proposed but it is yet to be seen if these are incorporated into the EPBC Act.

THE NATIONAL STRATEGY FOR THE CONSERVATION OF AUSTRALIA'S BIODIVERSITY

The major overarching policy at the national level is the NSCABD, finalised in 1996 by the Commonwealth and the states and territories after consultative development. The Strategy fulfils an obligation to develop a national strategy under the 1992 United Nations CBD. ANZECC (2001) reviewed the implementation of the Strategy. Such detailed reviews of major policies are crucially important to informing future policy and management but are not always carried out. The review of the Strategy forms an important basis for future monitoring of indicators that examines the response of government and society to the pressures on biodiversity. The *Implementation of the 1996 National Strategy for the Conservation of Australia's Biodiversity* box presents the summary of progress against the Strategy's objectives in the review.

Although the review of the NSCABD (ANZECC 2001) indicates considerable progress in implementing the Strategy, several issues were identified that, while the subject of existing efforts, need to be advanced over the next few years:

- managing key threatening processes - vegetation clearance, invasive species and dryland salinity
- avoiding future threats - identification of components of biodiversity and key threatening processes, identification and inclusion in management of ethnobiological knowledge, improving integrated management approaches and

bioregional planning, and raising public awareness of biodiversity.

Managing such key threatening processes and future threats will be critical if substantive improvements are to be made in biodiversity conservation in Australia.

IMPLEMENTATION OF THE 1996 NATIONAL STRATEGY FOR THE CONSERVATION OF AUSTRALIA'S BIODIVERSITY

Land clearing approval rates in New South Wales: 1998-2000

The clearing and removal of native vegetation in New South Wales is governed by the *Native Vegetation Conservation Act 1998* (NVC Act) in combination with the *Threatened Species Conservation Act 1995* and the environmental planning law, the *Environmental Planning and Assessment Act 1979*. The NVC Act has objects including the protection of 'native vegetation of high conservation value' and the prevention of 'inappropriate clearing of native vegetation'. These objects are to be achieved 'in accordance with the principles of ecologically sustainable development'.

The NVC Act provides a range of mechanisms for the preservation of native vegetation and for the approval of clearing. The general rule is that in order to clear native vegetation, the development consent must be sought and obtained from the Minister administering the NVC Act. 'Clearing' of native vegetation is defined extremely broadly. The likely effect of clearing on threatened species is assessed as part of the development consent process against a statutory eight-part test.

The NVC Act contains a mechanism for the making of regional vegetation management plans (RVMPs). These plans are intended to be the primary instruments for conservation and management of native vegetation. The rationale for regional planning is to take account of regional variations in vegetation and the specific issues that arise (e.g. clearance of 'woody weeds' in western New South

Wales). As of October 2000, no RVMPs have been formally made, although some are at an advanced stage of preparation.

The total area of native vegetation for which approval to clear was granted between January 1998 and August 2000 was 208 360 ha with the largest total amount of clearing approved in the North Coast Central West and Far West regions of the State. To date, most applications to clear vegetation under the NVC Act have received approval. Of the 1525 applications conclusively determined as at October 2000, 91.6% of applications were granted consent with conditions. Only 2.16% of these applications were refused, 1.57% were rejected prior to formal assessment and 4.65% were withdrawn by the applicants.

However, some clearing may not proceed despite successful applications for consent. Further, the total area approved for clearing may be less than the area that landholders originally sought to clear. Departmental documents indicate that in practice all clearing approvals are granted conditionally (DLWC 2001a). These conditions generally require the retention and replanting of trees within the approved area to ensure, among other things, preservation of habitat, and protection against land degradation (DLWC 1999; DLWC 2001a, 2001b).

Although approvals to clear are made subject to conditions, the recent (2000) decision of the Land and Environment Court in *Carr v. Minister for Land and Water Conservation* (2000, p. 175) has thrown some doubt on the Department's approach. The decision overturned a particular development consent on the grounds that specific conditions imposed on the consent to mitigate environmental effects of vegetation clearance had no valid basis in law.

The requirement of the NVC Act to seek consent to clear vegetation is subject to a range of exemptions. These exemptions include purposes ranging from 'private native forestry', 'minimal clearing' (i.e. clearing up to 2 ha per year) and clearing of regrowth less than ten years old. They also cover clearing in the Western Division of New South Wales, clearing on exempt land tenures (e.g.

national parks, state forests and listed wetlands), clearing in exempt local government areas and forms of clearing authorised under other legislation (e.g. *Rural Fires Act 1997*). In total, there are at least 37 situations in which clearing may be exempt from the operation of the NVC Act. As the Act contains no notification mechanism for clearing under exemptions, the extent of such clearing is not known or included in statistics showing total vegetation cleared.

According to the Minister for Agriculture (2000):

Some of the exemptions are poorly defined and open to interpretation. There are also concerns about the cumulative impacts of the exemptions and how they are being monitored and complied with.

The ambiguity surrounding some of the exemptions is causing difficulties in relation to enforcement of the NVC Act (DLWC 2000). In response, a formal review of exemptions was announced in late August 2000. An internal document (DLWC 2000) reviewing the exemptions states:

The exemptions which were meant to act as short term transitional arrangements have become the most problematical part of the Act in terms of achieving compliance.

Further, it says: 'The exemptions appear to be the most significant cause of inappropriate clearing' (DLWC 2000).

It appears there has been little activity regarding formal enforcement of the Act. There have been no prosecutions formally commenced and only nine stop work orders issued since January 1998. By October 2000, 112 warning letters had been sent in response to a total of 471 alleged (but not proven) breaches of the Act. Some observers may conclude that this approach to enforcement may be undermining the effectiveness of the Act in deterring inappropriate and unauthorised vegetation clearance.

In conclusion, the effectiveness of the New South Wales regime for control of native vegetation clearance has been affected by problems of ambiguous exemptions, a willingness to approve most clearing applications and an apparent reluctance to prosecute serious breaches of the NVC Act.

Albatross and Bycatch Policy

In 1995, data revealing an estimated take of 44 000 albatross annually through ensnarement on Japanese longlines set for tuna fisheries in the Southern Ocean gained considerable scientific political and public attention. Although applying to all seabirds, it was the threat in particular to vulnerable and endangered species such as the wandering, Amsterdam and northern royal albatross that led to the listing of longline fishing as a key threatening process under the Commonwealth government's [then] environmental legislation, the *Endangered Species Protection Act 1992*. The provisions of this Act have been retained under the new Commonwealth legislation, the EPBC Act. The original Schedule 3 listing triggered a requirement that a Threat Abatement Plan (TAP) for the incidental catch or bycatch of seabirds during oceanic longline fishing be created by Environment Australia. The TAP was to be a nationally coordinated action to alleviate the effect on seabirds of longline fishing in Australian waters.

Although statutorily required within a maximum of three years, the plan took slightly longer to complete. When released, its stated objective was to reduce seabird bycatch to below 0.05 seabirds per thousand hooks: a 90% reduction in bycatch. The TAP also maintained zero bycatch of seabirds in longline fisheries as an ultimate goal, especially endangered albatross and petrel species. The plan adopted two years ago is to be reviewed at the end of five years of operation.

Under the recommended TAP, a series of bycatch mitigation gear and method modifications have been required. These include the use of tori poles or bird scaring devices, the thawing of bait or weighting of lines to increase the sink rate, a prohibition on the discharge of offal from vessels (in particular during setting and hauling), and the practice of night setting when the birds are, in general, less active. There has not been any dedicated measure of bycatch reduction levels since adoption of the TAP, which makes it difficult to evaluate the efficacy of the policy intervention.

Although anecdotal evidence suggests that these modifications have been broadly successful, unnecessary delay and interagency disagreements have hampered the process. Given the limited five-year life of the plan, these delays may prove to be a significant impediment to the successful reduction of longline seabird bycatch. One area where disagreement has proved problematic is that of the level of observer coverage necessary, and the source of funding, to support such a program. Initial studies suggested between 10 and 15% on-board observer coverage, which was costed at more than $3.5 million over three years. Prolonged discussions in this regard have resulted in an interim agreement to trial ten underwater setting devices, gear technology which has been successful in reducing seabird bycatch in longline fisheries in other countries. The approximate $560 000 cost of these trials will be apportioned at 40:60 between industry and government, respectively. In addition, Australia has also taken the lead in negotiating a regional albatross conservation agreement under the auspice of the United Nations Convention on the Conservation of Migratory Species of Wild Animals, which was signed in June, 2001.

Monitoring and Evaluating Policy and Law

Evaluating biodiversity policy is a challenging task with the vast array of biodiversity and biodiversity-related laws, policies and programs. Evaluation can take several forms and operate at different resolutions. In this Report, the summary assessments of the application of vegetation clearance controls in New South Wales and implementation of albatross and bycatch policy are evaluations at a finer level. However, such evaluations are time consuming, and unless they have been carried out independently, or the required data are kept in a suitable manner and are easily accessible, it is not possible to conduct evaluations specifically for SoE reporting.

It is unlikely that detailed policy evaluation can ever be part of a broad scale national SoE Report, but the need certainly exists.

More difficult again is the evaluation of how well biodiversity issues are incorporated into other policy areas (e.g. transport, trade, regional development and taxation). Such evaluation is certainly outside the scope of SoE reporting and moreover is rarely undertaken in a detailed manner. Yet it is an important issue that governments should consider to ensure that indirect causes of biodiversity loss are attended to.

Biodiversity Issues and Challenges

This section considers each of the 65 biodiversity indicators outlined by Saunders et al. (1998). Where available, major supporting data, tables and figures are provided for these indicators, and details given for the key biodiversity issues and challenges facing Australia. The section is grouped into major areas: disturbance regimes and biodiversity, exotic species and GMOs and protecting biodiversity. Several key issues are examined in each major area and the relevant indicators explored.

Human Population Growth and Density

The relationship between human population and the environment, including biodiversity, is complex. The total environmental effect of a human population is a function of the nature of the environment or resource in question, which is highly variable, and a range of factors that determine the *per capita* environmental effect of each person (Dovers 1997b). These factors include resource consumption, waste production, recreational activities and other direct interactions (e.g. spread of weeds), management or regulatory regimes for land use and the remedial strategies in place.

Human populations in 1997 by IBRA region and projections for change in population density through to 2006 show that the pattern of human settlement in Australia is characterised by high rates of urbanisation, low-density cities and the concentration of the population within 50 km of the coast, mainly between Melbourne and Brisbane in south-east Australia and in south-west Australia. In

Antarctica, the total Australian population is around 300 people in summer and 75 in winter. The *Human Settlements* Report describes human population patterns in considerable detail. The focus here is on the potential effects of settlements on biodiversity.

Regions where population growth might increase pressure on biodiversity will depend on the resilience of regional environments and the management (or lack thereof) of the activities of the human population. Considering other indicators reported on in this report, pressures on biodiversity may be exacerbated by growth in human population in several coastal bioregions, especially in southern and eastern Australia. Population decline, such as is occurring in many Australian localities, may also influence biodiversity through a lessened ability to monitor or manage the environment.

Although the generation of population projections by IBRA is useful, population effects are not contained within such boundaries. Flows of people and the associated effects occur across bioregional (and other) boundaries, such as through both domestic and international tourism, the extraction or use of natural resource for consumers elsewhere (e.g. transport corridors or air or waterborne wastes). Such interregional flows of people and their effects on biodiversity point to the need for integrated approaches between different levels of government and other stakeholders, and of the need to incorporate biodiversity considerations into, for example, land use planning and transport policy.

Some people believe, on environmental grounds, that Australia's current human population (estimated as 19.4 million in August 2001) is too high. Others believe that the environmental effect of population growth can be managed. However, population is a complex, cross-sectoral issue and it is likely that the situation varies across different parts of Australia. Although population and settlement issues are managed to an extent through state, territory and local government planning, there is a strong case for national-scale coordination and development in order to consider the possible effects of population change for issues such as biodiversity.

CLEARING, FRAGMENTATION, DEGRADATION OF NATIVE VEGETATION OR MARINE HABITAT

The conservation of native vegetation is vital to biodiversity conservation. Vegetation is a key element of biodiversity in Australia since it comprises tens of thousands of plant species, thousands of vegetation communities and assemblages, and provides habitat to myriads of microorganisms and animal species. Native vegetation is also integral to the functioning of landscapes and ecosystems. As a consequence, the clearance and modification of native vegetation has several effects far beyond losing trees, shrubs, grasses and seagrasses.

Terrestrial Vegetation [BD Indicators 2.1 and 2.2]

European occupation of the Australian continent has resulted in significant changes in the extent and condition of native terrestrial vegetation. Clearance of vegetation for agriculture in the higher-rainfall regions, as well as those with more fertile soils and close to settlements, was promoted by governments. In all coastal regions of eastern Australia, this activity resulted in the removal and major modification of many vegetation communities including grassland, mallee and closed forest ecosystems. In some instances, the clearance of regional vegetation communities approaches 100%. Continental studies of landscape features such as biophysical naturalness (estimate of the extent to which regional plant and animal communities have been disturbed by modern technology), level of disturbance and wild rivers indicate that much of continental Australia has been highly modified by human activities, including areas that until recently have remained relatively inaccessible to Europeans.

The broad land use types in Australia have been categorised simply as two regions: the Intensive Land-use Zone (ILZ), and the Extensive Land-use Zone (ELZ). Typically, the nature and extent of vegetation clearance differs significantly between these two zones. Most Australians would be more familiar with the former that includes the intensive agriculture zone of the Murray-Darling Basin where the ecological and biological effects of clearance are increasingly self evident and irreversible. In contrast, the effects of vegetation modification on biodiversity in the ELZ is generally much less obvious to the human eye. Even today, some well-informed scientists and many members of the public think of northern Australia as 'pristine'.

THE EFFECTS OF VEGETATION CLEARANCE

The impact of broad-scale vegetation clearance on natural heritage and biodiversity is profound and has been of concern for several decades in Australia. The immediate effect of clearance of terrestrial native vegetation on plant and animal species can be significant. For vertebrate animals, comparative estimates of the population density of woodland birds indicate that between 1000 and 2000 birds permanently lose their habitat for every 100 ha of woodland cleared (Glanznig & Kennedy 2000), while it has been estimated that the clearing of mallee for wheat kills more than 85% of the resident reptiles (Glanznig & Kennedy 2000), on average, more than 200 individual reptiles per hectare. Longer-term effects of native vegetation clearance on species result from habitat loss and fragmentation combined with other threats.

The links between clearance of native vegetation and changes in hydrological cycles are relatively well understood and have serious implications for land management and biodiversity. Vegetation clearance changes the water balance of an area and this may lead to fundamental changes in the local soils and climate, as well as the local water table and its chemical composition (Stirzaker et al. 2000). Extensive clearance of vegetation across a catchment or region may generate 'cascading effects' (e.g. Mac Nally 1999) on the biophysical systems of the area and these changes may be irreversible or difficult to deal with other than through long-term (decade or century) mitigation and restoration strategies (Blackmore et al. 1999).

One recently publicised example of the physical changes that can occur to the hydrological cycle and environment as a result of vegetation clearance is dryland salinity. This has led to the recent development of several state and regional salinity strategies or audits. Western Australia developed their Salinity Action Plan in 1996, and Victoria, New South Wales and the Murray-Darling Basin Commission followed in 1999 to 2000. The most recent assessment of dryland salinity at the national scale was released in early 2001 by the NLWRA (2001). This assessment estimated that 630 000 ha of remnant native vegetation and associated ecosystems were within regions with areas mapped to be 'at risk'. These areas were projected to increase by up to 2 000 000 ha over the next 50 years. Dryland salinity poses a major threat to biodiversity in Australia.

In October 2000, the Commonwealth government released a National Action Plan for Salinity and Water Quality (Commonwealth of Australia 2000) to help manage dryland salinity and deteriorating water quality more systematically in key catchments and regions. The National Action Plan explicitly includes the conservation of biodiversity as one of its goals, to help ensure that the policy responses introduced to tackle land degradation are also consistent with the requirements for biodiversity conservation at the landscape level.

Salinity, Biodiversity Conservation and a New National Action Plan

Australia has critical salinity and water quality problems that require urgent attention. At least 2.5 million hectares (5% of cultivated land) are affected by dryland salinity (Commonwealth of Australia 2000) and this could rise to 17 million hectares at the current rate of increase in this type of land degradation (NLWRA 2001). In addition, one-third of Australian rivers are in extremely poor condition (e.g. within 20 years, Adelaide's drinking water is predicted to fail World Health Organization salinity standards in two days out of five). Infrastructure (e.g. buildings and roads) is being severely damaged in many rural urban centres.

Most salinity problems in Australia result from broad-scale land clearing, such as the type of land clearance occurring in Queensland.

The clearance of vegetation can fundamentally change the hydrology of an area, and lead to a significant increase in the volume of water draining beneath the normal root zone of the local vegetation. If stored salt is present in the soil or ground water, then hydrological changes may result in salt concentrations dramatically increasing in near-surface soils and/or at ground level.

Dryland salinity can have a serious effect on biodiversity. For example, preliminary findings from a four-year biological survey of the Western Australian agricultural zone indicate that 450 endemic plant species are under threat of extinction from salinity (CALM 2000a).

Further, the death of trees and shrubs in many wetlands in the wheat belt as a result of salinity has caused a 50% decrease in the number of waterbirds using them, and without intervention, about 75% of the waterbird species in the region will severely decline. If all wetlands in the wheat belt become saline, well over 200 aquatic invertebrate species will become regionally extinct (CALM 2000a).

In October 2000, 'A National Action Plan for Salinity and Water Quality in Australia' was released by the Commonwealth government. This plan proposes that concentrated action by governments and communities needs to lead to land use change supported by the application of scientific advances in mapping salinity, targeted tree planting and new cropping systems to manage salinity and water quality. The plan identifies high priority, immediate actions to address salinity, particularly dryland salinity, and deteriorating water quality in key catchments and regions across Australia.

The goal of the plan is to:

- prevent, stabilise and reverse trends in dryland salinity affecting the sustainability of production, the conservation of biodiversity and the viability of our infrastructure
- improve water quality and secure reliable allocations for human uses, industry and the environment.

The Action Plan builds on the work established under the NHT, the Murray Darling Basin Commission, state and territory strategies and the COAG Water Agreement. The plan will be implemented through targets and standards for natural resource management, particularly for water quality and salinity, including salinity, water quality and associated water flows, and stream and terrestrial biodiversity based on good science and economics. To ensure that integrated catchment or region management plans contribute to the achievement of nationally agreed outcomes, catchment or region-specific targets for salt, nutrients, water flow regimes, water quality, stream and terrestrial biodiversity will be required. This capability will need to be able to:

- map salinity hazard using 'ultrasound' technology and assess catchment and region conditions and issues
- maintain and improve the condition of existing native vegetation
- establish multiple purpose perennial vegetation (focused on agriculture, forests, biodiversity and carbon credits) in targeted areas, identified through salinity, vegetation and hydrology mapping, and ground water modelling
- protect and rehabilitate priority waterways, floodplains and wetlands
- improve environmental flows, where this is beneficial
- improve stream water quality using engineering works in critical areas (e.g. salt interception devices and ground water pumping, removal of weirs and redundant structures, fish ladders (to assist fish migrate upstream past structures such as dams) and artificial wetlands)
- install drainage in catchments or regions where agreed by affected land managers, the downstream effects are positive, and the overall benefits of the scheme provide substantial long-term results over other approaches
- address the harder adjustment and property amalgamation issues

- address the problems of degradation of rural urban infrastructure (e.g. buildings and roads).

There are some 20 high priority catchments and regions that need attention, including: the Burdekin-Fitzroy (Qld), Lockyer-Burnett-Mary (Qld), Namoi-Gwydir (NSW), Macquarie-Castlereagh (NSW), Murray (NSW), Goulburn-Broken (Vic.); Glenelg-Corangamite (Vic.), Midlands (Tas.), South-East (SA), Avon (WA), Northern Agricultural Region (WA), South West (WA) and Ord (WA-NT).

The plan acknowledges that land clearing in salinity risk areas is a primary cause of dryland salinity. It indicates that effective controls on land clearing are required in each jurisdiction, and states that:

- any Commonwealth investment in catchment or region plans will be contingent upon land clearing being prohibited in areas where it would lead to unacceptable land or water degradation
- the Commonwealth will require agreement from relevant states and territories (particularly Queensland, New South Wales and Tasmania) that their vegetation management regulations are effectively used or, where necessary, amended to combat salinity and water quality.

The National Action Plan is intended to promote major systemic improvements in land and water management. The Plan suggests that attention will need to be given to other high priority natural resource management issues such as the broader conservation of biodiversity and preventing productivity decline in other catchments and regions. Vegetation clearance also results in changes to the physical and chemical composition of soils and may significantly increase the likelihood of soil erosion and nutrient loss (AAS 2000), whereas clearance itself may lead to soil compaction and other physical modifications of the landscape. These changes may affect biodiversity deleteriously, both directly and indirectly (MDBC 1999; Stirzaker et al. 2000).

Vegetation Clearance

Concern about the fate of the nation's native vegetation and its associated biodiversity led to the adoption of two key targets in *National Strategy for the Conservation of Australia's Biodiversity* (1996) so that by the year 2000:

- Australia will have avoided or limited any further broad-scale clearance of native vegetation, consistent with ecologically sustainable management and bioregional planning, to those instances in which regional biodiversity objectives are not compromised
- Australia will have arrested and reversed the decline of native remnant vegetation.

As will become clear in the following sections, these targets are not close to being met.

National figures on vegetation clearance: The most widespread quantitative study of land cover change occurred in the ILZ that covers 38% of the continent (Barson et al. 2000). This study estimated that from 1990 to 1995, about 1 212 000 ha of woody vegetation were cleared for agriculture (cropping), grazing and other activities such as urban development. Although not all of this clearance may result in the permanent loss of woody vegetation, and does not include grasslands or sparse woodlands, these data are a compelling reminder of the spatial extent and intensity of vegetation clearance. These data also indicate that the relatively high rates of vegetation clearance recorded during the 1980s have continued into the 1990s. Recent data for 1997 to 1999 indicate that the rate and area of native vegetation cleared remains very high, especially in Queensland, and in some cases the rate of clearance has increased over the past three years.

Information about types of vegetation cleared and the nature of landscape modification and biodiversity depletion resulting from vegetation clearance is vital. The effects of native vegetation clearance on the extent of particular ecosystems and ecological communities are well known. For example, the estimated original extent of the

Big Scrub (subtropical rainforest between Lismore and Bangalow in northern NSW) was over 75 000 ha. However, by 1900, it had been reduced to about 300 ha scattered over 10 remnant patches. An assessment of areas in Queensland where the dominant vegetation type is Brigalow (*Acacia harpophylla*) shows that of the original estimated extent of more than six million hectares, only about 5% remains, and only about 30 260 ha, or 0.5%, were reserved. Most clearance occurred in the 1960s, and by the 1970s a large portion of the brigalow had disappeared (Nix 1994).

Glanznig and Kennedy (2000) reported that nominations for several biodiversity 'hotspots' that are threatened by vegetation clearance for sugar cane were being prepared by conservation organisations to inform governments of high priority areas for conservation protection. These hotspots include lowland forests, freshwater wetlands, grasslands, littoral rainforest and other ecosystems along the eastern seaboard bioregions of Queensland.

Barson et al. (2000) provided another indication of the changes in particular vegetation types in their review of land cover change in the Intensive Land-use Zone for 1990 to 1995. Using the J. Carnahan broad vegetation mapping as a baseline (Commonwealth of Australia 1990), these authors reported that open forest and woodland ecosystems represented much of the cleared area.

For example, 147 650 ha and 515 990 ha of these ecosystems, respectively, were cleared in Queensland between 1990 and 1995. About 350 000 ha of low woodland was also cleared across Australia during the same period, with some 90% occurring in Queensland (Barson et al. 2000). On a more positive note, this and related studies (e.g. Graetz et al. 1995) suggested that the extent of clearance of closed forests such as rainforest is lower than estimated previously.

At the end of the 1990s, the total area of native vegetation that has government sanction for clearing remains high, in excess of one million hectares per year. Rates of clearance varied across Australia, with Queensland estimated to have the highest rate of clearance:

about 425 000 ha of vegetation removed per year between 1997 and 1999 (Department of Natural Resources 2000).

Various estimates of clearing rates for New South Wales are 14 028 ha per year for 1997 to 2000 (Department of Land and Water Conservation 2001), 30 000 ha in 1999 (AGO 2001) and 100 000 ha in 2000 (ACF 2001). Based on estimates compiled by the ACF (2001), the total area of native vegetation that was cleared in Australia during 2000 was over 564 800 ha; the Australian Greenhouse Office (AGO) estimate for 1999 is 468 844 ha. On available figures, the former area is exceeded by only four other countries in the world: Brazil, Indonesia, DRC (Congo) and Bolivia.

PERMITS TO CLEAR NATIVE TERRESTRIAL VEGETATION

Governments are able to influence the rate of native vegetation clearance through enacting and enforcing legislation that requires permits to clear vegetation. There is no uniform legislation of this type in Australia. At the Commonwealth level, in early 2001, the Minister for the Environment, following advice from the Threatened Species Scientific Committee, listed land clearing as a key threatening process to biodiversity under the Commonwealth EPBC Act. In general, where legislation for land clearance and vegetation management is in place at the state and territory level, it covers woody native vegetation and provides for many exemptions. The legislation rarely applies to all land tenures and is politically controversial in that private property rights, government involvement, economic growth, regional development, land stewardship, Indigenous land rights, land access and biodiversity conservation are contested (Glanznig & Kennedy 2000). In addition, if the legislation is not enforced, it matters little how much is in place.

Number of Permits Granted [BD Indicator 18.1]

During 1999, Australian governments granted permits for clearing a total area of well over one million hectares of vegetation. The State governments of Queensland and New South Wales, alone, granted permits to clear 713 515 ha. Since the completion of SoE (1996), the total area of land for which vegetation clearance permits have been granted each year has generally remained at a relatively high level or increased. Selected examples of the number of permits granted are given below.

Queensland: Permits on leasehold land in Queensland increased from 496 957 ha on average between 1995 and 1997, to 644 515 ha in 1999. Recent data reveal that there has been a 31% increase in Queensland land clearing permits for the first six months of 2000 (431 781 ha), compared with 1999 (329 714 ha), and a 104% increase compared with the first six months of 1998 (211 199 ha). Old growth vegetation (bushland) comprises 166 194 ha of the 431 781 ha, with over 90 000 ha of this vegetation type approved for clearing in May 2000 alone.

New South Wales: In New South Wales, about 86 000 ha of native vegetation was approved for clearing in 1998, and at least 69 000 ha were approved in 1999. Although few verifiable, concise figures are available, about 40 000 to 50 000 ha per year are reported to be converted to improved pasture or feed crops for cattle, and 5000 to 6000 ha cleared for mixed horticulture (e.g. bananas and mangoes). Numerous proposals exist for major irrigated agriculture and horticulture projects of up to 250 000 ha and some commentators estimate that between 500 000 and one million hectares of native vegetation are earmarked for clearing over the next 10 years (ACF 2000).

Significant areas of native vegetation have been cleared in several regions of New South Wales that support depleted ecosystems. For example, the annual mean clearing rate for 1995 to 1997 in the Cobar Peneplain is estimated to be 13 250 ha. More than one-third of this region's native vegetation has been cleared since European settlement and <2% of the region is protected in the conservation estate. Other regions of major concern because of high annual rates of vegetation clearance, depletion since European settlement and poor reservation include the Brigalow Belt South, Darling Riverine Plains and North Coast.

Western Australia: Clearance of native vegetation in south-west Western Australia has been historically high, but is now under strict control. However, even low levels of clearance can have major effects on biodiversity, as it is likely to affect a greater proportion of the remaining vegetation. Effect on issues such as salinity, erosion and water quality are also of serious concern.

Tasmania: The Tasmanian government has made a commitment to best practice vegetation management and relies heavily on incentives and cooperation with landowners to achieve this, rather than regulation. On the basis of the available empirical data, this approach is grossly inadequate. The estimated average annual rate of clearance of native vegetation in this State between 1988 and 1994 is 10 429 ha. For 1994 to 1999, the mean rate of vegetation loss declined to slightly under 7000 hectares per year, and vegetation types of conservation significance continue to be cleared (Kirkpatrick & Mendel 2000). This brings the estimated total area of vegetation cleared in Tasmania for 1972 to 1999 (since consistent quantification commenced) to 265 575 ha. The large-scale clearing of native vegetation remains one of the most significant issues affecting Tasmania's natural environment (DPIWE 2000).

Northern Territory: Some 10 000 ha of vegetation are estimated to have been cleared in the Northern Territory during 1999, but precise details of the clearing are limited. Data for the Shire of Litchfield indicate the success of clearance applications. The Shire is about 310 000 ha in area, of which 27 000 ha are zoned as conservation, and removal of vegetation is subject to the Litchfield Area Plan 1992. The Plan requires that where removal of vegetation is proposed for an area exceeding 50% of an allotment, the vegetation be removed in accordance with environmental guidelines. Since 1996, requests to clear over 2500 ha of vegetation have been made and all appear to have been granted. The reasons for clearing have included horticultural production (mainly mangoes). Eucalyptus woodland (48%) and Eucalyptus open woodland (25%) were the predominant vegetation types cleared.

VEGETATION CLEARANCE: AN OVERVIEW

The estimates for rates of native vegetation clearance given above underestimate the true picture since they do not account for illegal clearing or clearing carried out under legislative exemptions (which covers regrowth and private forestry). In New South Wales, for example, such exemptions may include 'day-to-day farm management', and in Western Australia exemptions include clearing for urban development. Planning and environmental assessment processes are in place for many urban areas, but the expansion of residential areas is still of concern.

Current technical limitations associated with the use of satellite imagery also may introduce errors and inconsistencies in the estimates of vegetation clearance rates, as can the use of different definitions for vegetation communities. Of particular relevance, the continental-scale monitoring does not pick up clearance of native grasslands or areas where tree cover is sparse, including the large areas of southern Australia where individual scattered trees are a dominant, and very important part of the landscape.

The imprecision of data sets that are used to estimate rates of native vegetation clearance can be used by governments to delay action to protect biodiversity. Although not perfect, estimates of the rate of vegetation clearance are indicative of the high rate of clearance, its spatial extent and clearance 'hotspots'. The seriousness of the effects of these human activities on biodiversity is clear, and a far more comprehensive response by governments is required urgently.

The rates of vegetation clearance in Queensland have received particular attention. Although the total area of cleared land is extraordinary by Australian and global standards, the effects on biodiversity resulting from land clearance in States such as New South Wales and Western Australia may be as significant or even more significant. This is because the proportion of vegetation cleared versus the amount remaining might be more significant on biodiversity than the absolute amount cleared. Measures to control land clearance in States such as New South Wales, Western Australia and Tasmania are, therefore, essential to minimise further degradation of the biodiversity associated with the vegetation in these jurisdictions.

New remote sensing techniques, a standardised approach to determining land cover change and further evaluation of methods will undoubtedly improve the precision of current estimates. While these improvements are now actively pursued by government agencies and research institutions, it is very telling that Australia is unable to systematically report on the rate of clearance by vegetation type at the national scale.

VEGETATION MODIFICATION AND FRAGMENTATION

McIntyre and Hobbs (1999) developed a framework describing the range of landscape alteration 'states'. Four landscape alteration states are recognised:

- intact
- variegated
- fragmented
- relictual.

These are associated with increasing amount of habitat destruction and decreasing levels of habitat connectivity. In intact landscapes (e.g. arid rangelands), less than 10% of the vegetation is destroyed and the landscape mosaic is, therefore, 'habitat' in various states of modification. At the other extreme are relictual landscapes (e.g. cropping or urban areas) where over 90% of the vegetation is destroyed and small areas must survive in a landscape matrix which may be hostile to the continued persistence of the vegetation.

The 'intact' landscapes described by McIntyre and Hobbs (1999) largely coincide with the ELZ, which covers around 60% of monsoonal, semi-arid and arid Australia. New data suggest that the relatively contiguous extent of overstorey vegetation in the ELZ, compared with that for the highly cleared ILZ, is not a reliable indicator of the conservation status of ecosystems. The disproportionate extinction and regional loss of mammal species for the arid zone and related parts of the ELZ is well known. However, recent and ongoing biological surveys suggest that pervasive changes in these ecosystems (e.g. fire regimes and livestock grazing intensity) that are closely linked to land tenure threaten a diverse range of biota (e.g. Franklin 1999; Fraser 2000; Woinarski 2000). In most instances, the conservation status of native vegetation communities is poor, especially in most parts of the rangelands, arid and wet-dry tropics.

In the ILZ (mainly southern and eastern Australia), large areas of vegetation and its associated biota have been heavily modified and they generally fall into the fragmented and relictual categories of McIntyre and Hobbs (1999). The clearance of native vegetation disrupts ecosystems and habitats and results in the creation of remnant islands or fragmented patches (e.g. near Jandakot airfield, WA) and linear fragments along roadsides. These have become important reservoirs of plant and animal species that depend on native habitats. The degree of vegetation modification and fragmentation in these regions and beyond has intensified during the past five years.

Few of the vegetation remnants remaining after clearing are large enough to sustain ecological processes such as water and nutrient cycling at the rates that existed before disruption. Many continue to be disturbed by threatening processes such as invasion by weeds or feral animals coming from the surrounding cleared land and firewood collection.

More insidious threats like rising saline water tables and the inflow of fertilisers from surrounding lands are also of major concern. Despite the major threat that dryland salinity poses to native vegetation (see *The effects of vegetation clearance* and *Salinity, biodiversity conservation and a new National Action Plan*), mapping of the distribution of major fragmented vegetation types in selected catchments and their likely response to projected hydrological changes has begun only recently. The issue of clearing by 'ecological action' is also critical. Putting a fence around patches of bush and keeping grazing animals out may not be sufficient to prevent further degradation, and in many cases loss, of remnants. Changes in the condition of remnant vegetation, such as the presence or absence of plant regeneration, therefore need to be assessed to guide and improve management.

THE NATURE OF FRAGMENTED VEGETATION

The destruction and modification of native vegetation has left a legacy of patches of native vegetation of various sizes, shapes, connectivity and condition. Many of the ecological values associated with native vegetation (Kirkpatrick & Gilfedder 1999; Lambeck 1999) relate to medium to larger patches, although all native vegetation has some role in the landscape (Williams 2000). For example, individual trees provide shade for stock, nesting and foraging sites for wildlife (Lumsden & Bennett 2000), cycle nutrients, act as a source of seeds and may help to reduce ground water recharge and to recycle cations from depth (Reid & Landsberg 2000). The gradual decline in vigour and eventual death of many of these trees has led to a phenomenon known as 'rural dieback'. There are several different causes of this widespread condition and their relative effects can vary in different parts of the farm, on neighbouring farms, in adjacent districts and catchments and between regions (Reid & Landsberg 2000). These authors have listed several factors that may cause rural dieback including insect damage, large numbers of Noisy Miners, secondary salinisation, pathogens, drought, nutritional disorders and old age. Most of these are made worse, or are associated with, the broad-scale clearing in the areas where rural dieback generally occurs.

BURNING THE BUSH: THE IMPLICATIONS OF FIREWOOD COLLECTION FOR BIODIVERSITY CONSERVATION

Most of the firewood supply in Australia comes from stands of remnant native vegetation on private property (Driscoll et al. 2000; Wall 2000). Firewood collection includes the removal of fallen and standing dead trees from the bush, as well as living trees that are sometimes ringbarked for future use. Overall, the harvesting of wood for domestic heaters means that around five and a half million tonnes per year is burnt, a similar volume to the amount of eucalypt woodchips exported each year (ANZECC 2000c ; Williams 2000). Firewood is, therefore, the third largest source of energy used in Australia after electricity and gas. Around 60% of firewood is purchased through small collectors or suppliers rather than firewood merchants with established premises.

The removal of such large amounts of dead and living wood from patches of bush is considered to have a major effect on the whole spectrum of biodiversity, from ecosystems to genes (Driscoll et al. 2000). Fallen timber provides habitat for insects and other invertebrates, reptiles and ground feeding birds such as the Bush Stone-curlew (Traill 2000). Dead, standing timbers, also targeted by firewood cutters, are more likely to have hollows than live trees and are favoured as nesting sites by possums, parrots, bats and other wildlife. It is also a favoured foraging site for some insects and insect-eating species such as the Brush-tailed Phascogale (Traill 2000). The potential loss of highly specialised species of invertebrates and fungi associated with coarse woody debris is of particular concern (Driscoll et al. 2000). The disappearance of these species from native bushland could affect ecosystem services such as nutrient cycling and plant establishment.

Inland forests and woodlands in lower rainfall zones appear to be the ecological communities most threatened by the collection of firewood (Driscoll et al. 2000). In Victoria, 49 ecological communities have been listed as potentially threatened by firewood collection, emphasising the extent of the problem. At the species level, firewood removal has been implicated in the decline of birds at the local and regional level (see Driscoll et al. 2000), while at the national level, Garnett and Crowley (2000) identified 21 bird species threatened by firewood collection. Plants can also be affected. In Tasmania, 13 species have been listed as threatened by firewood collection. Many of these species have very restricted distributions and it is thought that firewood collectors could inadvertently damage a large proportion of the remaining populations of these plants. The spread of fungi such as *Phytophthora cinnamomi* by firewood collectors is also considered a threat (Driscoll et al. 2000).

Recently at the national level, government and NGOs have paid attention to firewood collection and its effects. For example, in mid-1999, the Victorian National Parks Association held the first national conference on firewood collection, and ANZECC is developing a national approach to firewood collection and use (ANZECC 2000c). The objectives of this strategy are to:

1. protect remnant native vegetation, threatened ecosystems and habitat for threatened and declining wildlife species
2. encourage ecologically sustainable firewood collection from native forest, woodlands and plantations
3. contribute to broader environmental objectives (e.g. improved air quality, dryland salinity, and contributing to carbon sequestration).

The codes of practice proposed for the firewood industry would be voluntary but given the scale of the industry and its significant impact on biodiversity, more concerted and urgent measures are required.

Fragmentation of native vegetation creates new edges between remnants and cleared or disturbed land which leads to 'edge' effects. These include physical changes to the remnant in the border region such as different levels of exposure to the sun and wind and changes in water cycles and the local air temperature (Saunders & Hobbs 1991). Biotic changes include invasion by opportunistic species with good dispersal or colonising abilities such as weeds and feral animals. Fragmentation also isolates and creates barriers between patches of native bush. In most cases, recently isolated remnants can be expected to continue losing species (Saunders & Hobbs 1991). The loss of a population of a species (that has declined to a size that is not viable) may take considerable time if individuals are relatively long lived. For example, it may take several hundred years to lose species of long-lived trees, particularly since adult plants are often less sensitive to changed environmental conditions than plants in their seedling and juvenile stages. This phenomenon also applies to many fauna (e.g. Trapdoor Spiders in the wheat belt of Western Australia may live for at least 23 years (Main 1999)).

The consequences of habitat fragmentation on biodiversity depend on the interaction of many factors that may vary for different species and habitats. To illustrate the implications of habitat fragmentation on a particular group of species, the effects of habitat fragmentation on birds is summarised.

Threats are divided into those that are continuing and those that no longer occur, either because the taxa are extinct or because the process has ceased and is no longer affecting the surviving birds.

Table : Threats to Australian birds

Clearance and fragmentation of habitat	Current threats		Former threats		Total
	Confirmed	Speculative	Confirmed	Speculative	
Agriculture	32	4	22	9	67
Mineral extraction	2	4	3	-	9
Softwood plantations	2	1	2	-	5
Urban development	4	3	1	-	8
Forestry operations	3	14	-	-	17
Total	43	26	28	9	106

Source: Glanznig and Kennedy (2000).

OTHER IMPACTS ON NATIVE TERRESTRIAL VEGETATION AND ASSOCIATED BIODIVERSITY

Significant modification of native vegetation may result from agents of change other than clearing such as grazing by introduced livestock and native herbivores, changes to the hydrological regimes leading to waterlogging and salinity and altered environmental flows, invasion by weeds and changes in fire regimes. Another threatening process that is receiving increasing attention is the effect of firewood harvesting on remnant native vegetation with current rates of extraction estimated at around 5.5 million tonnes per year, similar to the amount of wood that is exported as woodchips.

The pressures on biodiversity in old growth forests were identified as a major issue in SoE (1996). At the time, the logging of native eucalypt forests was receiving considerable attention in the national media, especially since the level of wood-chipping of native forests had been very high throughout the decade and many ecologists were seriously concerned about the effects of intensive forestry practices on forest biodiversity and ecosystem services. For example, Norton (1996) argued that many forestry practices in eucalypt forests in eastern Australia were not ecologically sustainable, and that biologically significant forest ecosystems and many native forest biota were threatened. One response of governments to attempt to resolve the debate over native forest management was to initiate the RFA process. RFAs arose from the ESD process of the early 1990s and were intended, put simply, to take care of the reasonable conservation needs of the forests, and then to facilitate economic development in the remaining forests (Kirkpatrick 1998).

Unfortunately, the RFAs do not provide a comprehensive coverage of the native forest estate as there are important areas that have not been assessed. Further, within the regions where RFAs were undertaken, many important conservation needs have not been adequately addressed. For example, several biologically significant ecosystems and species have not been adequately protected, many additions to the conservation reserve network have not been determined using the best available scientific techniques, and the efficacy of a number of forestry management prescriptions remains to be determined (e.g. Kirkpatrick 1998). The implications of these limitations for biodiversity conservation may be amplified since government quotas on wood-chipping were removed on signing of an RFA. Hence, the potential for the intensification of wood-chipping in these regions on public and private lands has significantly increased.

THE LIVING LANDSCAPES PROJECT: A COMMUNITY BASED PROJECT TO DEVELOP SUSTAINABLE LANDSCAPE MANAGEMENT

The Living Landscapes Project is a community-oriented research and strategy development project in the wheat belt of south-west Western Australia, which is attempting to embed biodiversity conservation into catchment and agricultural planning. This pilot planning process considers both agricultural production and broader landscape issues such as nature conservation and ecological health (Frost et al. 2000). The aim is to assist community groups to develop landscape management practices that protect biodiversity within an ecologically viable and sustainable land use system. The Living Landscapes project involved an interdisciplinary team that used experiential learning as an overarching process (Frost et al. 2000). The other key process used was the focal species approach (Lambeck 1999). By combining these two approaches, Living Landscapes has developed a set of guiding principles for nature conservation planning in the context of sustainable land management. The approach is now being considered in several regions in eastern Australia.

THE RELATIONSHIP BETWEEN HABITAT FRAGMENTATION AND BIRD ABUNDANCE AND RANGE: A CASE STUDY

Studies have investigated the medium-to long-term effects of fragmentation on different groups of species, in particular, birds and mammals. In relation to birds. The threats are divided into those that are continuing and those that no longer occur, either because the taxa are extinct or because the process has ceased and is no longer affecting the surviving birds. For instance, the clearance of mallee in the Western Australian wheat belt has now almost stopped but the effects on fragmented populations are continuing.

Numerous studies in many States have concluded that bird abundances are directly related to the degree of habitat loss and fragmentation, and heavily fragmented areas will be accompanied by net losses of species which can continue long after the initial clearing

(Recher & Lim 1990). In Western Australia, Saunders (1989) found that a rapid loss of species had occurred in wheat belt reserves since the clearance of the original vegetation 30 to 50 years ago. This finding has been reinforced by a more recent study which showed that 49% (95 of the 195 species) of birds recorded in the wheat belt (excluding vagrants) have declined in range and/or abundance since the region was developed for agriculture. Most of these losses result from loss of habitat and fragmentation of the remainder. This general pattern of regional loss and decline of bird species has been repeated in other States such as New South Wales (Reid 2000a, 2000b) and Victoria (Bennett & Ford 1997).

The degradation of habitat by removal of the understorey of forest and woodland ecosystems, through grazing for example, is also significant since it can simplify these ecosystems and result in the loss of species and genetic variability. For example, a study in the Latrobe Valley of south-east Victoria found that small, heavily grazed patches (less than 10 ha) supported fewer forest birds and an increased number of farmland birds, including Noisy Miners (*Manorina melanocephala*), which aggressively excluded other species. Few birds ate insects in the canopy of these patches, which showed signs of dieback due to insect damage. Planting understorey species may be the most effective way to exclude Noisy Miners and encourage other native bird species.

Recher (1999) predicted that if immediate action was not taken to reverse the decline of native birds, then Australia would lose half of its terrestrial bird species in the next century. The most urgent actions identified were to end the clearing of native vegetation, remove inappropriate fire regimes, control feral and native animals whose abundance threaten native species and restore functional ecosystems.

Effects of grazing on vegetation and biodiversity in Australian semi-arid and arid rangelands have been quantified (Landsberg et al. 1999) using artificial sources of water (tanks and dams, bores, waterholes and wells) as a measure of potential effect by livestock.

Most large mammals require regular access to drinking water, and in arid environments its availability determines where they graze. Thus, sources of water become foci of grazing activity. This may result in a zone of accentuated impact around each water point where the vegetation is browsed and perhaps killed, soils are compacted and habitat for various flora and fauna is modified or destroyed. CSIRO Sustainable Ecosystems has mapped the location of artificial sources of water to evaluate the potential effect on biodiversity, and has concluded that vast areas of the rangelands previously beyond the reach of large grazing animals may now be exposed to sustained grazing pressure. Further, few potential reference areas for determining pregrazing patterns of biodiversity now remain.

Surveys of water points for the Biograze project revealed that the effect of grazing by cattle is minimal beyond 9 km and beyond 6 km for sheep. Therefore, the map is a conservative estimate of the impact of grazing for sheep. The water point data are accurate for rangeland regions in central, northern and western Australia. Areas closer to the coast, especially in south-west and south-east Australia, have too many water points to be mapped. Consequently the figures provided in these regions are an underestimate. The areas masked out of the analysis had incomplete datasets.

Declines in temperate woodland ecosystems arise from proximate factors such as population reduction and species extinction, genetic loss, substrate modification and salinity (Sivertsen & Clarke 2000). Clearing for grazing and grazing of the understorey of these ecosystems by introduced livestock and feral animals have been identified as important threats (e.g. Kirkpatrick & Gilfedder 2000; Landsberg 2000; Lunt & Bennett 2000). Typically, grazing and related human-driven disturbances exacerbate the deleterious effects on biodiversity resulting from the extensive clearance of lowland woodland ecosystems. These types of degradation o: ecosystems have been conceptualised as part of an underlying slow process of desertification (Bauer & Goldney 2000). In general, as grazing pressure increases, native species become less abundant and are replaced by exotic species (Yates et al. 2000). However, under certain circumstances, grazing animals do not need to be totally excluded from native vegetation and in some instances the presence of grazing has been associated with the maintenance of high conservation values at a site (Williams 2000).

Regrowth Vegetation

Conservation of vegetation regrowth is vital for biodiversity conservation. Regrowth is important because it may provide habitat for key elements of biodiversity that have been affected by vegetation clearance and fragmentation. Regrowth is also important in supporting and sustaining biophysical and ecological processes. For example, as vegetation regrows, the structural, floristic and biological composition of areas change and may reduce the extreme nature of habitat fragmentation for resident species, and favourably modify local climate and environmental regimes (e.g. energy, radiation, light and exposure to extremes in ambient temperature) (Saunders & Hobbs 1991). The notion that regrowth vegetation has little or no value for biodiversity seems widespread, and one consequence of this is that regrowth clearance may proceed without adequate planning, or may not be challenged even when it threatens significant components of biodiversity (Kirkpatrick & Gilfedder 1999).

No net loss: The policy of 'no net loss' of native vegetation (e.g. Bushcare's target within Australia by July 2001, as per Department of the Environment and Heritage - Portfolio Budget Statement 2000-2001), and the associated policy of offsets (to negate the negative effect of clearing by separate actions that have positive effects), are being considered by several jurisdictions. There are many issues associated with these policies that could have long-term effects on biodiversity. For example, replacing mature woodland with an equivalent area of saplings could satisfy the no net loss criteria, but these two vegetation types are very different in structure and function. Once the original vegetation disappears from a site, then it is difficult, if not impossible, to recreate it. And although revegetation projects are becoming increasingly sophisticated, it will take decades to develop the characteristics of the original vegetation (i.e. being self sustaining), especially the benefits provided by large trees. These not only provide hollows on which many Australian birds and mammals depend, but they produce a more reliable and extensive source of nectar for a range of fauna (Wilson & Bennett 1999). Consequently, there is general agreement that the first step to sustainable management is to retain existing native vegetation where possible (Williams 2000). The next steps are to protect and manage that vegetation and then, where appropriate, to revegetate cleared areas.

Offset actions for native vegetation clearing could include improving the management of existing native vegetation, revegetating a previously cleared area or establishing a tree plantation. Biodiversity conservation would be undermined by these types of activities if exemptions (e.g. vegetation of high conservation could not offset and thus be cleared) were not comprehensive and legally enforceable. State management agencies in New South Wales are attempting to benchmark the concept of biodiversity credits (biodiversity values in major ecological communities such as woodlands, rangelands and forests) to provide a basis for trading.

In 1997, the Victorian government released Victoria's Biodiversity Strategy. One important goal of this strategy package is to achieve a reversal, across the entire landscape, of the long-term decline in the extent and quality of native vegetation. This is designed to lead to a 'net gain' in vegetation with the first target being no net loss by 2001 (The State of Victoria 1997).

In this case, the goal of no net loss was based on the premise that (although 'natural' is best) it is possible to recover the extent and quality of native vegetation by active management intervention (The State of Victoria 1997). Questions relevant to this goal include whether the plant species in the vegetation patch are locally native, whether are there enough of the locally native species remaining in a patch to warrant inclusion of the patch in accounting for native vegetation and when revegetation could be classified as part of the State's native vegetation estate.

Restoring vegetation: Tree replanting schemes and similar activities are now supported by government, industry and community. However, frequently the amount of vegetation replaced may be orders of magnitude less than the loss of plants as a result of clearing, modification, dieback and other threatening processes. For example, the goal of the Bushcare program, which is the largest in the NHT, is to reverse the long-term decline in the quality and extent of Australia's native vegetation cover. This ambitious goal will not be met, at least in the short term, because of the continuing broad-scale clearing in areas such as Queensland and New South Wales.

Paton et al. (2000) reviewed the distribution, status and threats of woodland ecosystems in South Australia and suggest that future revegetation strategies in the State need to include a greater diversity of plants, use locally endemic species and plant these in natural dispersion patterns to maximise biodiversity benefits.

The removal of extensive areas of deep-rooted perennial native vegetation and replacement by shallow-rooted annual crops and pastures has significantly affected the hydrological cycles of many regions. As a result, the water table that was formerly drained by the deep-rooted vegetation is rising to the soil surface at rates of up to 0.5 m per year and at least 2.5 million hectares are affected by dryland salinity. Up to 80% of some catchments in Western Australia, for example, may need to be replanted with trees to reverse the rising salinity, and some salinised streams may already be beyond recovery in this region (Glanznig & Kennedy 2000). Clearly, where revegetation and restoration is possible, catchment-based strategies need to be consistent with biodiversity conservation.

Vegetation restoration is a relatively young science and until recently in Australia has focused on ecosystems and vegetation communities where mining and intensive forestry operations have been undertaken. Restoration of native vegetation can be attempted by active (e.g. planting and soil inoculation strategies) or passive (e.g. natural regeneration) means. While the ability to restore a cleared vegetation community or highly modified community to its original state in terms of species composition and ecosystem function is largely untested, substantial progress has been made in the development of techniques to restore some components of the biota to mine sites (e.g. Nabalco, Gove bauxite mine in northern Australia).

Area Cleared or Modified to Area Revegetated *[BD Indicator 18.2]*

Increases in woody vegetation (woody plants greater than 2 m in height and 20% cover) in Australia's more intensively used agricultural areas for 1990 to 1995 were found to be 414 000 ha (excluding that due to regeneration following fire and native forest harvesting) (Barson et al. 2000).

The estimated loss of woody vegetation for the same period was 1 282 200, including clearing for agriculture, mainly cropping (213 680 ha), grazing (929 280 ha), on-farm tree planting (1 760 ha) and plantation management (90 160 ha). This shows a net loss of native woody vegetation for the period.

Recent work on landscape design principles and guidelines is attempting to enhance the rehabilitation and restoration of (at least) terrestrial landscapes. Examples of guidelines and principles are the perceived thresholds for vegetation cover (i.e. a minimum of 30% tree cover is being recommended in temperate and subtropical regions not experiencing severe salinity) (see Barrett 2000a; Williams 2000) and the focal species approach being used for revegetation of cleared agricultural landscapes (Lambeck 1999).

Plantations: The practice of commercial plantation forestry is expanding across many regions of southern Australia. A recent study based on computer modelling of 12 regions by the Australian Bureau of Agriculture and Resource Economics (ABARE) suggests that some 19 million hectares of cleared agricultural land may be suitable for commercial plantations (Burns et al. 1999). Potential expansions of this scale may have significant implications for regional development and infrastructure, and biodiversity conservation. For example, the preparation of land for plantation establishment may involve a significant input of fertiliser and the use of chemicals to control weeds and poisons to control native vertebrate herbivores. These inputs can directly and indirectly affect biota.

Aquatic Systems

Vegetation clearance, vegetation modification and intensive land use also affect aquatic plant ecosystems and freshwater fisheries. Bunn et al. (1999) demonstrated that native vegetation that overhangs streams and rivers is important in moderating water temperature, which in turn affects native fish and other biota.

Acid sulfate soil poses another potential threat to biodiversity. These soils are rich in iron sulfide and are common in coastal areas. They are relatively stable while saturated with water, but the sulfide forms sulfate and sulfuric acid when drained and exposed to air. The sulfuric acid can kill fish, prawns, oysters and other aquatic life when washed into waterways by rain. Iron sulfate soils on which vegetation is regrowing can also cause problems once they become exposed. The Coastal Acid Sulfate Soils Program, which is managed by Environment Australia through the NHT, has provided funding for projects with onground works that demonstrate better options for managing coastal acid sulfate soils.

By area, most stream catchments have been subjected to medium to high levels of disturbance, including significant portions of the Gulf of Carpentaria Drainage Basin, Lake Eyre Drainage Basin and Bulloo Drainage Basin that support a relatively low human population density.

Aquaculture also affects aquatic plant ecosystems and related parts of the catchments used for this purpose. Data on the current and potential effects of aquaculture on biodiversity are limited and disparate. The extent of aquatic systems for which cultivation permits have been granted are increasing and suggest this issue needs further investigation. For example, of the 567 lakes in western New South Wales, 70 (74 136 ha) have government permits for cultivation.

Extent of Each Vegetation Type within Protected Areas *[BD Indicator 13.1]*

In a Statement on the Environment by the Australian Prime Minister in 1992 (Commonwealth of Australia 1992c), it was announced that 'The Government has adopted a policy that all ecosystems be surveyed and that a comprehensive, adequate and representative system of reserves be established progressively by the year 2000'. The National Reserve System Program was designed to deliver this outcome.

The NRSP and related state and territory programs, the RFAs, the Indigenous Protected Areas (IPA) scheme of Environment Australia, new multitenure management schemes, and the enormous growth in contributions from the non-government sector (e.g. Trust for Nature and Bush Heritage Fund) have helped to increase the spectrum of the nation's system of conservation reserves. Even so, many anomalies exist and very few regions in Australia support a conservation reserve system that meets all of the requirements of comprehensiveness, adequacy and representativeness (CAR).

The representation of major vegetation types within the Australian conservation reserve estate remains poor despite the long-standing recognition of the need to enhance the reservation and protection of these ecosystems.

In Queensland, for example, native vegetation is typically underrepresented in the conservation estate, including in those areas (e.g. around Charleville and Emerald) where rates of land clearing are very high by international standards and threats to biodiversity are extraordinarily high. Sattler and Williams (1999) reported that over half of the 287 regional ecosystems in the Mulga Lands, Desert Uplands and Brigalow Belt bioregions in Queensland are endangered or threatened with extinction.

Only 63% of these endangered and threatened ecosystems are represented in the conservation reserve system of these bioregions and less than one-third of these ecosystems is found in more than one protected area.

New South Wales is one of the more 'data-rich' parts of Australia, although its detailed biological data sets, like others elsewhere, are localised. To overcome this limitation, Pressey et al. (2000) have developed a new classification of landscapes at a scale of 1:250 000 across the whole 802 000 square kilometres of New South Wales. The classification is derived mainly from abiotic data (e.g. topography, soils and climate) and, in conjunction with new data on native vegetation cover, has allowed the first quantitative state-wide review of protected areas and future priorities at a scale that approaches decisions about land use. Pressey et al. (2000) found that most of the 1486 landscapes in New South Wales are poorly reserved relative to an indicative conservation target of 15% of the total area of each.

In the eastern 60% of New South Wales, gaps in the reserve system are related to the concentration of reserves on land with high ruggedness and low potential for intensive land use. Pressey et al. (2000) mapped the relative priority of landscapes to indicate the urgency of conservation action to prevent conservation targets being compromised (or further compromised) by clearing of native vegetation. Mapping of priorities shows large differences within and between natural regions and land tenures. More than 9% of private land is occupied by high-priority native vegetation and, across the whole State, about 85% of high priority vegetation occurs on private land. This indicates the importance of controlling vegetation clearance on private land if biodiversity effects are to be avoided.

Biodiversity Effects of Clearing, Fragmenting and Degrading Marine Habitat [BD Indicator 2.1]

Australia's marine environment extends from the shores and wetlands along the coastline to the abyssal deep, and from the coral reefs of Torres Strait in the north to the pack ice of the antarctic continent in the south. The warm tropical waters off northern Australia are only 1000 km from the equator while the nation's coolest waters are some 3700 km to the south (Zann 1995). From east to west, the coastline of the Australian continent spans a distance of almost 4000 km, from Cape Byron to Shark Bay. The marine environments include extensive, well-developed and specialised reef systems, giant kelp forests, seagrass meadows, mangrove ecosystems and near-shore systems, and sand and mud-bottomed habitats that cover much of the continental shelf.

Effects on Marine and Coastal Biodiversity *[BD Indicators 2.1 and 2.2]*

Several human activities destroy or significantly modify and degrade marine habitat and its associated biodiversity. These threatening activities and processes include eutrophication and pollution from elevated levels of nutrients, dumping of wastes including ballast water, overfishing and collecting, incidental bycatch, introduction of exotic organisms, loss of habitat, bioaccumulation of noxious materials like heavy metals, pesticides, herbicides, siltation, tourism, oil spills, downstream effects of dams and dykes and fishing litter. The destruction of habitat on the seabed has many parallels to habitat clearance on land, with sedimentation on the continental shelf being considered one of the most important factors in the decline of marine biodiversity.

Estuaries: Estuaries are the meeting place of fresh and salt waters. Naturally rich in nutrients, estuaries are ecologically highly productive, and provide habitat for fish, migratory birds and other species. In Australia, estuaries and sheltered bays have also been the focus of urban and industrial development and recreation.

Australia has 783 major estuaries: 415 in the tropics, 170 in the subtropics, and 198 in temperate areas (Zann 1995). A recent national evaluation has shown that the geographical extent of relatively undisturbed estuaries continues to decline. The long arid coastlines in the south-west and west have few estuaries. Most river catchments in eastern and southern Australia have been extensively cleared. This has resulted in land erosion, sedimentation of rivers, and increased sedimentation and levels of nutrients downstream in estuaries, bays and adjacent coastal waters. High sediment levels in the water reduce light penetration, which affects rates of photosynthesis. When sediments settle they can also smother seabed organisms. Sedimentation of estuaries and shipping channels causes shoaling and alters currents. Sedimentation is a major problem in ports and shipping channels, necessitating regular dredging. This resuspends sediments, creating further environmental problems.

POLLUTION SOURCES ON THE GREAT BARRIER REEF CATCHMENT

Human population densities in north Queensland are low and concentrated along the coastline. Only five cities have populations exceeding 40 000. Grazing of cattle for beef production is the largest single land use (77%) in the catchment with cropping, mainly of sugar cane (3%), and urban and residential development (3%) considerably smaller in areal extent. Other significant land uses include mining (coal and metalliferous) and cotton cropping.

Beef cattle numbers are about 4 500 000 with the highest numbers in the Fitzroy catchment. The sugar cane cultivation area has increased steadily over the last 100 years with a total of 390 000 ha reached by 1997. Other industries with significant expanding land use (and fertiliser use) are cotton (mostly in the Fitzroy catchment) and horticulture (in many catchments), particularly bananas.

Overall, 66% of the estimated nutrient and sediment flux is estimated to come from grazing lands, with 8% from cropping lands and 26% from 'pristine' areas. The total run-off flux of sediment is estimated to be four times the amount prior to European settlement.

Water quality: Where rivers drain disturbed acid soils, such as in northern New South Wales and southern Queensland, estuaries may become acidic periodically. This increases levels of dissolved aluminium and iron, which form compounds very toxic to fish. As a result, fish diseases such as 'red-spot' disease and fish kills are increasingly common in estuaries. At least 64% of estuaries in New South Wales and 22% in Victoria are considered to have poor water quality. Elevated nutrients in river systems may cause eutrophication, and the excessive growth of algae can smother other organisms and deplete oxygen levels. Catchments in the tropical north are less affected by human activities, but may carry heavy sediment loads resulting from soil erosion. Unnaturally elevated sediments and nutrients have led to major declines in seagrass beds in temperate Australia. Poor water quality and loss of habitat have also caused a decline in estuarine fisheries. For example, fisheries are thought to be threatened in 21% of estuaries in New South Wales and 23% in Victoria. However, eutrophication of some estuaries has enhanced their value for oyster aquaculture (Barratt et al. 2001).

Coastal lakes and lagoons: Of great concern in south-east and south-west Australia is the declining water quality and eutrophication of coastal lakes and lagoons and the effects these conditions have on biodiversity. Settings of particular concern are those that are insufficiently flushed by the sea, such as Tuggerah Lakes and Lake Macquarie (NSW), Gippsland Lakes (Vic.) and the Peel-Harvey system (WA). As coastal lakes are largely restricted to the densely inhabited south-east coastal strip, a significant proportion of Australia's coastal lakes have been degraded (Zann 1995). Conservation of these areas is essential as they may support habitat and feeding grounds for water and shore birds including migratory birds such as Whimbrel, Eastern Curlew, Terns and Sandpipers.

Intertidal shores: The intertidal shores are the meeting place of the land and sea. Shores are periodically immersed by sea and exposed to air, and thus subject to extremes in salinity. They are often enriched by land nutrients and have a high biological productivity. Shores have a high diversity of very specialised animals and plants, and may require special conservation measures. Australia's shores include open coasts with rocky headlands, cliffs and sandy beaches; and sheltered coasts, bays and estuaries with muddy and sandy tidal flats. Intertidal rocky shore habitats are often limited in area. They are also potentially vulnerable to human impacts. Threats to shore communities include over harvesting of molluscs, crustaceans and sea urchins for food and bait, trampling by fishers and other visitors, oil slicks and other pollutants which float on the sea surface, and loss of habitat. In the more populous south-east, south and south-west of Australia, significant areas of shores around coastal cities and towns have been reclaimed or alienated by sea walls, port development, industry, housing and tourism, and recreational facilities (SoE 1996).

Coastal disturbance: A summary (Zeller 2000) of disturbances in Queensland resulting from human structures and activities (e.g. water diversion, dam building, disturbance of spawning habitats, increase in the number of sewage treatment plants, gravel and sand extraction, and port, resort and real estate developments) showed that during the 1990s, disturbance of fisheries habitats has generally increased as human activities have expanded and intensified along the coastline. For example, waste water generated by about 2.4 million people is treated by sewage treatment plants in the coastal zone of Queensland. In excess of 90% of the treatment plants located within 50 km of the coastline discharge into coastal freshwaters, estuaries or inshore marine waters. These discharges have been treated to a 'secondary stage' where most suspended solids and bacteria have been removed. However, few of these discharges have their nutrients removed.

Commercial channel dredging for sand and gravel occurred from around the turn of the 1900s in the Brisbane River estuary and probably in excess of one million cubic metres were removed during the 1990s. In 1996, 390 000 m^3 of sand was approved for extraction from northern Moreton Bay. From 1990 to 1996, about 2467 ha of tidal land supporting marine plants were reported as being disturbed by human activities, particularly for port expansion, tourist developments and marina/boat harbours.

Salt marshes: Australia has some 13 595 square kilometres of salt marshes (Zann 1995). Coastal salt marshes are intertidal plant communities dominated by herbs and low shrubs, and are often associated with estuaries. Salt marshes are highly productive, and support key habitats for many organisms including migratory species such as the rare Orange Bellied Parrot (*Neophema chrysogaster*) in

Victoria. A major threat to salt marshes in built-up areas is land reclamation. Extensive areas have been filled for ports, marinas, canal estates and urban and industrial sites. Other threats include degradation by rubbish dumps, off-road vehicles, invasion by weeds (particularly by introduced cord grass, pampas grass, para grass and rushes), periodic surges associated with low-pressure systems and drainage for mosquito and sandfly control. Human developments that result in the total loss of salt marshes have typically occurred in south-east Australia, where biodiversity and endemism are highest.

Mangroves: Mangroves are tree and shrub species that are adapted to the periodically inundated and salty conditions between the tides. Mangrove forests are very productive ecosystems and are of major ecological and economic importance. They provide habitats and nurseries for many fish, form a buffer for estuaries from sediments and for coastlines from storm waves, are natural nutrient filters, and are critical habitats for many birds and other wildlife. Australia has the third largest area of mangroves in the world, and has some of the most diverse communities, but these are poorly protected from habitat destruction and degradation.

Mangrove ecosystems, in general, exhibit a decline in the diversity of species from the tropics to temperate zones. Similarly, within the tropics, there is a decline in diversity of mangrove species from wet tropical environments to the arid tropics. At a finer scale again, mangroves in the Great Barrier Reef show a general decrease in diversity with increasing latitude and a cross-shelf decrease from the mainland coast to the islands.

Overall losses of mangroves in Australia are small compared with those from other countries. However, locally significant losses have occurred around Australian coastal cities and towns. For example, about 20% of mangrove ecosystems have been cleared in Moreton Bay, near Brisbane, for coastal development. Elsewhere, in areas such as the Hawkesbury and Parramatta Rivers in Sydney, it is thought that accelerated silting associated with poor land management has led to significant expansion of mangroves. The major threats to mangrove ecosystems include continued local clearing and development, and the effects of various human-induced catchment alterations. Only some 8% of Australia's mangrove communities are in protected areas. Protection may be given through other avenues (e.g. in Queensland, mangroves are protected under Fisheries legislation).

Seagrasses: Some of the world's largest seagrass habitats occur in Australian waters. Seagrasses are marine flowering plants and 30 of the world's 58 species are found in Australia (SoE 1996). Seagrass beds are ecologically important because of their high productivity, their ability to trap and stabilise sediments, their importance as fisheries habitats, and as the habitat for important species such as Dugongs and turtles. Australia has the highest biodiversity of seagrasses in the world, the largest areas of temperate seagrass and one of the largest areas of tropical seagrass. Anthropogenic changes that are degrading these habitats and threatening biodiversity include eutrophication, heavy metals and toxins, changes in hydrology, sediment run-off, mining and dredging, trawling, moorings and boat propellers and introduced species (Zann 1995).

Australia's unique temperate seagrass beds appear to be under particular threat. Increased sedimentation and nutrients from catchments have been linked with massive dieback of seagrasses in many areas. New South Wales has lost half of the *Zostera* seagrass in its estuaries. In Victoria, around 85% of the total biomass of seagrass in Western Port Bay has been lost. In Tasmania, there have been declines in the Hobart and D'Entrecasteaux region, Triabunna and St Helens on the east coast, and Tamar, Port Sorell and Duck Bay on the north. In South Australia's Gulf of St Vincent, around 5000 ha of seagrass has been lost, and this trend is continuing (Government of South Australia, Environment Protection Agency 2000). Some regrowth of seagrass is occurring in the Port Adelaide area, but its survival is uncertain. In Western Australia, around 97% of seagrass in Cockburn Sound have been lost. A serious loss of tropical seagrasses has occurred in Hervey Bay, Qld, causing major mortality of Dugongs.

Long-term monitoring of seagrass beds by the CSIRO in a study area located in the north-west Torres Strait revealed a 30% loss of seagrass between 1989 and 1993 (Long et al. 1997), and this trend apparently continues. An estimated 1119 square kilometres of seagrass were lost in the north-east region of the study area. Combined, the net losses that have been quantified in this area alone represented a 10% reduction in the estimated total area of seagrass in the Torres Strait (Long et al. 1997). Once lost, seagrasses do not readily recover and this has serious implications for the species these ecosystems support.

Coral and rocky reefs: Coral reefs are among the most productive, diverse and complex ecosystems in the world and support a significant proportion of the planet's marine biodiversity (Zann 1995). Coral reefs are also under global threat.

A global assessment of reefs (Wilkinson 2000) shows continuing decline with 27% of the world's reefs having been effectively lost, with the largest single cause being the massive, climate-related coral bleaching event of 1998, which destroyed about 16% of the world's coral reefs in nine months.

Climate change adds a new dimension to threats to reef ecosystems since changes in ocean temperature, ocean currents and flows of nutrients, disturbances regimes and sea level rise have the potential to destroy many reef ecosystems. Because of the general decline in many coral reefs, the biodiversity conservation values of Australia's reefs are of growing importance, and the management of reef ecosystems is attracting even greater attention (Zann 1995; Wachenfeld et al. 1998).

Australia has the largest area of coral reefs of any nation and the largest coral reef complex, the Great Barrier Reef (Zann 1995; Wachenfeld et al. 1998). Major areas of coral reefs are also present in Torres Strait, the Coral Sea Territories, and central and northern Western Australia. The Oceanic Shoals Bioregion, which consists mainly of submerged reefs, is also a very significant coral reef province, with many unexplored reefs on the Sahul Shelf and along the edge of the Timor Trough. The Tasman Sea reefs (Elizabeth and Middleton Reefs and the Lord Howe Island fringing reef) are the highest in latitude in the world, and thrive in conditions otherwise marginal for coral growth. High latitude reefs, such as the Abrolhos Islands, are also found on the west coast of the continent.

Of the reefs that have been well studied, a striking feature of coral reefs is the natural variation which can exist in reefs in close proximity to each other.

General issues affecting Australia's coral reefs include effects of sediments, agricultural chemicals and nutrients, effects of fishing and tourism, the threats of oil spills, and negative changes in habitats as a result of enhanced climate variability and climate change (Zann 1995; Wachenfeld et al. 1998).

Specific threats include elevated nutrients in the inner Great Barrier Reef, outbreaks of Crown-of-Thorns Starfish (*Acanthaster planci*) in the outer central and northern Great Barrier Reef and Tasman reefs, damage from the passage of tropical cyclones, and outbreaks of coral-eating *Drupella* snails in Ningaloo Reef, WA. Wachenfeld et al. (1998) summarised the major environmental attributes on the Great Barrier Reef and the state, pressure and responses associated with them.

Coral reefs are relatively well represented in Marine Protected Areas in Australia, although only a small area is protected in zones where fishing is prohibited. Because of the general decline in many of the world's coral reefs, tourism values of Australia's reefs are growing.

Tourism on the Great Barrier Reef is actively managed, which should keep habitat destruction and degradation to a minimum. However, the annual increase in visitation rates and increased mobility of visitors to coral reefs as a result of new and improved forms of transport is likely to have some effect. The increased mobility of tourists is an important management issue for all marine environments, including shallow waters amenable to underwater tourism (Wachenfeld et al. 1998).

NATURAL VARIATION OF BIODIVERSITY IN CORAL REEFS

Results from three, separate long-term studies of corals have emphasised variability of reefs through time. In one 30-year study on Heron Island in the southern Great Barrier Reef, coral cover was found to vary between 0 and 80% in different patterns depending on the site.

In another study covering six reefs in the central and northern Great Barrier Reef between 1980 and 1995, substantial changes occurred in the numbers of areas dominated by corals, bare substrate or other organisms, but by the end of the study, the proportions were similar to those at the beginning. However, this overall similarity in average conditions masked that some reefs had improved while others had degraded. At one inshore reef, an area once dominated by macroalgae was replaced by a coral-dominated community.

4

Unexplored Ecosystems: Seamounts and Terrestrial Cliffs

Many ecosystems in Australia remain virtually unexplored because they are inaccessible, or because the species they support are cryptic or inconspicuous. Two such areas are oceanic seamounts, and the cliffs of the terrestrial environment.

In June 2000, CSIRO reported that hundreds of new species have been discovered on seamounts in the Coral Sea and the Tasman Sea between New Caledonia and Tasmania. The south-west Pacific contains the greatest density of seamounts and seamount ridge systems in the world. Each seamount is an ecological island and many different species have evolved independently in each area. The communities on seamounts are dominated by corals, sponges, sea fans and other organisms that filter their prey from the strong currents characteristic of this environment. There are about 30 000 seamounts in the world's oceans, but most species previously known from this deep ocean environment come from sampling only five of these. The study of 25 seamounts uncovered 850 species, more than previously reported from all studies of seamounts in the past 125 years. About one-third of these species are new to science and are likely to be restricted to the seamount environment.

Cliff environments are a unique combination of substrate and physical conditions that support unique biotas that depend only partly on mass effects from adjacent communities. Many are relatively undisturbed by human activities. Larson et al. (2000) reported on the substantial undersampling of cliff faces in terrestrial environments. The discovery of the Wollemi Pine (*Wollemia nobilis*) in a sandstone gorge in the Blue Mountains is symptomatic of this sampling bias. This very conspicuous species remained undiscovered close to Australia's largest human population for 200 years because there have been no studies of the vascular or non-vascular flora of the associated cliffs.

There is little doubt that many new species and unique communities remain to be discovered in cliff environments, on oceanic seamounts and in many other unique ecosystems.

Despite their high conservation and economic values, Australia's temperate reefs are inadequately studied scientifically, and relatively few are protected. Although very little is known of the effects of human activities on temperate reef habitats, it is often assumed that outside metropolitan and industrial areas, they are relatively unaffected (Zann 1995).

The continental shelf and sea floor communities: Continental Australia has around 2.5 million square kilometres of geomorphic continental shelf, half of which is less than 50 m deep. The continental slope, which drops from a depth of 150 m to 4000 m, has an area of at least two million square kilometres (Zann 1995). An extensive continental shelf is also found adjacent to the AAT and there is much to be discovered about the biodiversity in this area.

Little is known of the effects of humans on sea floor communities, although seamount ecosystems have attracted study. The rate of sedimentation on the sea floor has greatly increased since European colonisation of Australia by factors of ten to 100, and even more in some areas.

Dumping of wastes on the sea floor is controlled under international conventions (e.g. the London Convention) and various Commonwealth and state/territory legislation. However, dredging and dumping of wastes may still cause localised disturbances around ports.

Trawling [BD Indicator 2.2] : Some shelf and slope fish species have been severely overfished, and trawl nets may dislodge attached species such as sponges and modify the habitat and food chains. Barratt et al. (2001) reported that the South East Trawl Fishery is increasing in intensity of trawling, and that the Twofold Shelf Region may be the most disturbed of the Interim Marine and Coastal Regionalisation for Australia (IMCRA) regions in south-east Australia. Trawling intensity and the area in the Northern Prawn Trawl fishery decreased substantially during the 1990s, but appear to be slowly increasing.

Trawling is of serious concern for certain marine biodiversity and recommendations have been made to list prawn trawling as a threatening process to turtles. For example, about 3% of the Loggerhead Turtle (*Caretta caretta)* population that forages in the Northern Prawn Fishery is killed annually as a result of this fishing activity (Barratt et al. 2001). Trawling can also have considerable effect on invertebrate benthos, although these organisms are often ignored in bycatch studies, which focus on retained bycatch. Work on fish trawling in Western Australia, however, has indicated that each pass of the trawl removed 16% of the large (>20 cm) macrobenthos, with multiple passes effectively denuding the substrate. This is a marine form of habitat fragmentation. Trawling should, therefore, be considered for its effect on the entire range of species, not just target and non-target fish caught.

Trawling is prohibited in some areas (e.g. in parts of the Great Barrier Reef Marine Park). Changes in the design of trawl nets used to catch prawns should reduce the chances of catching unintended species such as cod and turtles. This will help to minimise unintended changes in species composition and trophic interactions in reef ecosystems.

The trawl net in the upper diagram has no bycatch reduction devices fitted. All animals that enter the net are caught in the cod end, including prawns, turtles and unwanted fish species. The trawl net in the lower diagram has two types of bycatch reduction devices fitted. The turtle excluder device prevents turtles from swimming into the cod end, forcing them out through a flap on the trawl net. Fish have a tendency to swim against the current and may swim out through the top of the trawl net. Prawns, however, are still caught in the cod end.

Hydrocarbon exploration: Levels of hydrocarbon exploration of the seabed and production activity tends to fluctuate markedly annually, and there has been a relatively small increase in both types of activity since the early 1990s. Petroleum industry activities and facilities occupy a small surface area of the seabed and are concentrated almost entirely in the region of the Gippsland Shelf and Carnarvon Shelf. Six rigs were in operation at September 1999 and these drilled almost 100 wells with a total length of over 280 km. Seismic surveys and drilling affect the biodiversity of benthic marine environments. Seismic surveying involves the introduction of a high-energy acoustic pulse into the earth's crust, such as the release of a high pressure air blast into the water, dynamite, electromechanical vibrations and steam injection. Sonic echoes from the sea floor are then received by a cable, up to one nautical mile long (1852 m), which is filled with hydrophones and is dragged by the surveying ship. In 1998, the total length of 2-D (a single line or track) and 3-D (systematic grid coverage of a discrete area) surveys exceeded 900 line kilometres.

Seamounts

Seamounts are remnants of extinct volcanoes, typically cone-shaped, 20 to 500 m high and several kilometres across at their base. About 70 seamounts arise from water depths of between 1000 and 2000 m on the continental slope, between 50 and 170 km off southern Tasmania (Tasmanian Seamounts). This field of seamounts is a distinctive geological feature not known elsewhere on the continental margin of Australia. They support a distinct benthic (bottom-dwelling) community of animals, many of which are native to the Tasmanian Seamounts and do not occur elsewhere. A high proportion of species (24-43%) new to science has been discovered, including at least eight new genera. The species in the area appear to be representative of the seamount fauna of the Seamount Region and have been protected recently by the Commonwealth government (Environment Australia 2000). Generally, however, most of Australia's sea floor is not actively managed.

Protected Area Management [BD Indicator 13.2]

Modern standards of best practice reserve management include the preparation and implementation of management plans or strategies that are well researched, developed in consultation with users, implemented, monitored and evaluated. Although the existence of a management plan does not describe the adequacy of that plan, the proportion of areas subject to a management plan is an important and basic indicator of the sophistication of protected area management within a jurisdiction.

PROTECTED AREA PROGRAMS

The NRSP, commenced in 1996, funds the acquisition of land as well as tying funding for new protected areas to the preparation of a management plan. ANZECC (2001) stated that 'management plans for major protected areas [funded under the Program] should be in place by the end of 2000'. Another focus of the Program is the development and application of 'best practice' standards of management. The Marine Protected Areas Program has a similar focus, and in addition, the development and management of the national Marine Protected Areas estate will be informed by an overall Strategic Plan of Action. The IPAP has a similar focus. All three programs are funded under the NHT.

MARINE PARKS AND RESERVES

There are 13 marine parks and reserves managed by the Commonwealth government and six protected areas with a marine component. Five of these have been declared since the SoE (1996): the Great Australian Bight Marine Park (April 1998), Tasmanian Seamounts Marine Reserve (May 1999), Macquarie Island (October

1999), Lord Howe Island Marine Park (June 2000) and Cartier Island Marine Reserve (June 2000). These marine parks range from 167 square kilometres (Cartier Island) to 19 769 square kilometres (Great Australian Bight) so represent a major addition to the marine conservation estate. In addition, the Heard Island Wilderness Reserve was established in 1996. Some state and territory governments have established, or plan to establish, marine reserves. For example, Western Australia has declared six marine parks and reserves with more expected to follow.

Significant marine areas are protected in all of the places where reefs are found, although management is often constrained by their great size and remoteness, and by lack of financial resources. The number of marine areas that are protected is small compared with the recognised diversity of marine environments, and the total area protected (outside of the Great Barrier Reef complex) is very small compared with that under Australia's jurisdiction.

The first IPA was proclaimed in 1998 in the Flinders Ranges region of South Australia on 'Nantawarrina', a property owned by the Adnyamathanha people. As at March 2001 there were 13 formally declared IPAs in Australia covering an area of around three million hectares: Dhimurru IPA (NT), Warul Kawa IPA (Torres Strait), Deen Maar IPA (Vic.), Nantawarrina IPA (SA), Yalata IPA (SA), Oyster Cove IPA (Tas.), Risdon Cove IPA (Tas.), Preminghana IPA (Tas.), Mt Chappell Island IPA (Tas.) and Badger Island IPA (Tas.), Watarru and Walalkara IPAs (SA), Guanaba IPA (Qld). Also, the Mutawintji National Park joint management agreement between Indigenous owners and the NSW National Parks and Wildlife Service (NPWS) was supported through the Program.The benefits of establishing IPAs have been well documented and are not only related to enhanced conservation of biodiversity but also social, educational and other aspects (e.g. De Lacey 1994; Thackway et al. 1996; Centre for Environmental Management 1999; Szabo & Chester 2000). These include:

- the preservation and/or enhancement of biodiversity by Indigenous peoples over many thousands of years in Australia is considered testimony to their superior ability over European practices to conserve nature

- management of protected areas by Indigenous peoples can be a very cost effective solution to conserve biodiversity for governments.

THE INDIGENOUS PROTECTED AREA PROGRAM

The impetus for establishing Indigenous Protected Areas (IPAs) stems from the potential benefits that such arrangements can have in maintaining or enhancing biodiversity through the use of traditional management techniques and that many areas of unique conservation status under the IBRA system are found only on lands owned or leased by Indigenous peoples. The Indigenous Protected Areas Program (IPAP) receives funding from the NHT and was initiated in 1995, the same year that the NHT was set up. It was not until 1997, however, that IPAP began in earnest. The NHT defines an IPA as being 'governed by the continuing responsibilities of Indigenous peoples to care for and protect lands and waters for present and future generations'.

The IPAP provides for two different forms of Indigenous involvement in protected area management:

- IPAs, where the establishment and management of protected areas is on Indigenous owned estates
- Cooperative Management, which involves the establishment of cooperative (joint) management arrangements over government-owned protected areas between Indigenous groups and the relevant government nature conservation agencies (Centre for Environmental Management 1999).

Number of Protected Areas with Management Plans *[BD Indicator 13.2]*

Although a significant proportion of protected areas are subject to management plans, clearly there are gaps in coverage and management planning is an ongoing and unfinished task.

It has not been possible to access historical data for this Report to show the pattern of increase or decrease in management planning activity. Wescott (1995) recorded the production of management plans for national parks (not other reserve categories) for Victoria in 1993 as follows: approved plan or strategy 20%, draft plan or strategy 26%, in preparation 21%, 'old' plan or strategy 10%, and without plan or strategy 23%.

In the 1990s, the conservation reserve estate has been expanded significantly in several jurisdictions in terms of gross area, number of reserves and the variety of ecosystems represented. Although Australia's conservation reserve system remains inadequate, this recent expansion is a positive and important advance compared to the status of protected areas in many parts of the world.

Information is not readily available to determine if the management of reserves is adequate to protect biodiversity, or indeed cater for human use, in the long term.

Integrated Bioregional Planning [BD Indicator 12]

Bioregional planning has been a relatively recent policy and planning response to the myriad of land and water management issues that Australia faces. What precisely constitutes 'integrated bioregional planning ' is unclear and probably will remain so, partly because of the varying contexts where the concept is implemented in different jurisdictions.

However, the core principle is that biodiversity occurs and functions across spatial scales that rarely match traditional political, administrative and other boundaries and functions, and that this feature needs to be explicitly catered for in policy, planning and management. The features of biodiversity are reflected in the NSCABD which includes the objective to 'manage biodiversity on a regional basis, using natural boundaries to facilitate the integration of conservation and production-oriented management techniques'. This objective has been assessed as 'partially achieved' (ANZECC 2001). The incorporation of biodiversity into an environmental management system (EMS) at both the bioregional and farm scale is starting to receive greater attention (Anderson et al. 2001).

Biogeographical Regions

A key to making progress with bioregional planning is to have an agreed system of regionalisation, which has been achieved in recent years. Identification of biogeographical regions is well advanced across Australia through IBRA (Thackway & Cresswell 1995) and IMCRA (Thackway & Cresswell 1996) processes. IBRA was developed in the mid-1990s to support the NRSP, and has become well established as a land-planning framework. IMCRA, which addresses marine and coastal areas, was published in June 1998. Several recent initiatives advance capacities for bioregional planning and these fall into two groups: those directly concerned with biodiversity, and a much larger array of other 'regional' arrangements which may or may not integrate biodiversity issues.

REGIONAL PLANNING DIRECTLY CONCERNED WITH BIODIVERSITY

Significant examples of this category, indicating trends in planning approaches, are outlined under the Commonwealth and various states as follows.

Commonwealth:

- The EPBC Act contains provision for the Commonwealth Minister to prepare bioregional plans for any Commonwealth area, and a requirement for the Minister to take account of such a plan in decision making.
- The cooperative management of the Great Barrier Reef Marine Park by the Queensland and Commonwealth governments has been achieved since 1975 and stands as the most long-standing and most internationally recognised integrated marine conservation approach.

Western Australia:

- The State government, through the NRSP has continued a series of regional biological surveys.

- In late 2000, a large-scale integrated regional research and development and planning program was established, covering the Ord River catchment and Bonaparte Gulf in northern Western Australia. The Ord Bonaparte Program (OBP) will combine analysis of marine, coastal and terrestrial components to integrate biophysical data-gathering and analysis with social, economic and institutional research and development. The aim is to develop regional resource management options to inform future decision-making.

Queensland: A basis for future planning and management has been established with the publication of *Conservation Status of Queensland's Bioregional Ecosystems* (Sattler & Williams 1999).

South Australia: A biodiversity plan for the south-east region has been prepared, and draft plans are under development for other regions.

Victoria: Victoria's Biodiversity, the state strategy, uses bioregions to assess and plan for nature conservation and threatened species and ecosystems, and to integrate these issues in land and water planning. Regional biodiversity strategies are being prepared and the first comprehensive draft strategy, for the Goldfields Bioregion, was released for discussion in August 2000 (Ahern et al. 2001). Victoria seems likely to be the first Australian jurisdiction to substantively connect bioregional and catchment planning and management, and this attempted coordination represents an important opportunity to monitor success and challenges.

New South Wales:

- Regional vegetation management committees have been established and RVMPs are under development for a range of areas across the state (as required under the *Native Vegetation Conservation Act 1997*).
- The NSW Biodiversity Strategy of 1999 lists bioregional assessment and planning as a priority. Several bioregional assessments are underway, such as the Darling Riverine Plains Project. This Project highlights the important role of

the community in making decisions that will assist in the conservation of natural and cultural heritage.

- Conservation assessments for a series of IBRA regions in New South Wales, for example the NSW Riverina and the Cobar Peneplain, have also been funded under the NRSP to help determine priorities for protection.

Northern Territory: Bioregional conservation planning has been undertaken in the Finke, Daly Basin and Sturt Plateau regions.

Australian Capital Territory: Regional SoE reporting, including biodiversity, is being developed by the Australian Capital Territory and 17 regional local government bodies.

Multijurisdictional planning:

- Coordinated management of protected areas and biodiversity in the Australian Alps is enabled by a Memorandum of Understanding between the Commonwealth, New South Wales, Victorian and Australian Capital Territory governments. This is effected mostly through interaction at the regional and district level by operational staff, and is regarded as a leading international example of effective conservation planning and management across jurisdictions.
- Increasing integrative regional planning endeavours in the coastal zone in several jurisdictions. Examples are the regional coastal management plans specified under the *Coastal Protection and Management Act 1995* (Qld), and the preparation of six coastal management strategies in Tasmania. Such integration of development and conservation was a major focus of recommendations arising from the Resources Assessment Commission's Coastal Zone inquiry.

OTHER REGIONAL INITIATIVES

This category encompasses a large range of regional initiatives that have arisen in recent years. These encompass social, environmental and economic issues and vary greatly in their structure, style of operation, issue and sectoral coverage, and stage of progress.

A hallmark of new regional initiatives is their inclusion of, or more often initiation by, community groups - unlike most of the more biodiversity specific initiatives above. A comprehensive review of emerging regional arrangements, including those examples mentioned here, was provided by Dore and Woodhill (1999).

Although the primary focus of many of these arrangements is regional economic development or land and water management, many include biodiversity. Examples include the Cape York Peninsula Land Use Strategy (CYPLUS), the Blackwood Catchment Initiative and the Lake Eyre Basin Management Group. The lack of government support for the associated Cape York Heads of Agreement, that was negotiated between pastoral, mining, Indigenous and conservation interests, diminished what many saw as a crucial precedent in negotiated regional strategic planning that had incorporated production, cultural and biodiversity issues. The most long-standing and influential regional assessment and planning process that takes biodiversity into account was implemented by Victoria's LCC. The LCC provided the informational and consultative underpinning for Victoria's reserve estate. The LCC was reconstituted in a different form under the *Environment Conservation Council Act 1997*.

The RFA process involved assessment and conservation area planning for biodiversity and production values in publicly owned native forests at regional (but not bioregional) scale. The RFA process involved the development of conservation criteria for biodiversity values. Unfortunately, the development and application of the criteria for reservation targets have not been consistent with the stated objectives of the RFAs and therefore have not resulted in the protection of forests with high biological values or the development of a CAR reserve system at the regional level.

Most land and water planning exercises have biodiversity implications, whether explicit or implicit, and increasingly these are developed at regional scales (including catchments). For example, the Macquarie Marshes Land and Water Management Plan (Brock 1997) covers environmental water flows and vegetation and fauna conservation. Provision of environment flows, which have to be considered under the COAG water reforms, will influence both instream and riparian biodiversity (Fisher 2000).

DISTURBANCE REGIMES AND BIODIVERSITY

Harvesting

Harvesting of native wildlife [BD Indicators 8.1 and 8.2]

The harvesting of native flora and fauna for domestic and export purposes is controlled by various legislation in each state and territory. The extent of native flora and fauna harvesting taking place in various states and territories is outlined in the following sections.

STATE AND TERRITORY-BASED POLICIES AND PROGRAMS

South Australia

In South Australia, information is collected only for the number of permits issued for the harvesting of native flora. Total permits issued for harvesting native flora have been increasing since 1996 when there were 36 issued. In 1997, 46 permits were issued, 1998, 56 and in 1999, 67.

For native fauna, data are kept on the number of permits issued. There are 98 permits issued to kangaroo shooters, eight permits issued to kangaroo processors and 815 permits (predominantly to property owners) issued to destroy protected animals. Similarly, there are 58 Emu (*Dromaius novaehollandiae)* farming permits issued and 57 permits were issued to people who wish to keep and sell eggs of protected species (e.g. emu eggs may be collected by Indigenous people for carving and can attract high prices on domestic and overseas markets).

Australian Capital Territory

The only data available on native fauna harvesting (by calendar year) for the Australian Capital Territory indicates that the trade of native fauna is growing, with a steady increase occurring in each category every year since 1996.

Tasmania

Data on native animal and plant harvest are limited and records are only kept for the harvesting of Brushtail Possum (*Trichosurus vulpecula)*, wallaby (e.g. *Macropus* sp.) and muttonbirds (*Puffinus* sp.), with annual values of production of around $400 000, $750 000 and $425 000, respectively. This industry provides a source of income for Indigenous communities in Tasmania. Statistics for the commercial harvesting of Brushtail Possum are available but these data do not document the non-commercial shooting of possums for crop protection purposes. For example, permit returns indicate that an estimated 246 158 possums were taken between 1 July 1998 and 31 June 1999 and another 156 410 between 1 July 1999 and 23 May 2000. Harvesting statistics are also provided for the Short-tailed Shearwater (*Puffinus tenuirostis*).

Northern Territory

The main native fauna harvested in the Northern Territory are kangaroos and crocodiles, with crocodile harvesting mainly carried out on farms. Since 1971, crocodile species have been protected and only 16 adult crocodiles have been harvested from the wild (in 1997) since this time. Some adult and juvenile crocodiles have been taken from the wild for restocking, but over the last two decades this has totalled only 67. The Parks and Wildlife Commission (PWC) of the Northern Territory remove 'problem' crocodiles from native habitats and these animals are then used for breeding purposes on farms or sold for meat or skins. During the 1990s, between 130 and 240 crocodiles have been captured under the problem crocodile program in the Northern Territory (PWC of the Northern Territory 1998). The farms have worldwide sales of crocodile leather, skins, flesh and other crocodile products.

The main form of harvest from the wild is for crocodile eggs that are used to stock crocodile farms. Provisions for harvesting eggs, hatchlings and subadults are made in management programs. The total number of eggs collected from the wild has increased from 2320 in 1984 to around 20 000 from 1997 to 2000. Since 1993 to 1997, the number of farms, skins and meat produced from these farms has steadily increased.

Western Australia

Since 1993, detailed records have been kept for harvest data of native flora in Western Australia. These data indicate no clear trend in the total number of stems harvested from 1993 to 1999. The harvest of native species from Western Australia is carried out predominantly for the export market and therefore often follows styles and fashions required by other markets. Harvest of individual species can fluctuate from year to year.

Successful captive breeding of the Australian Saltwater Crocodile (*Crocodylus porosus*) has occurred on crocodile farms in Western Australia and Johnston's Crocodile (*C. johnstoni*) has been bred at farms in Broome and Wyndham. During 1994, 639 *Crocodylus porosus* were bred in captivity, with a further 746 in 1995. As in the Northern Territory, the taking of crocodiles from the wild is subject to quota limitations and since the mid-1980s, the numbers taken have fluctuated. The rate of harvesting depends on the requirements for restocking and also the climatic conditions that may limit the supply of eggs and young. Similarly, there has been no discernible trend in the number of crocodiles processed under licence for the seven years to 1998.

New South Wales

The harvesting of kangaroos in New South Wales is undertaken to use the full carcass (as opposed to skin-only shooting). Since 1996, New South Wales has harvested a higher percentage of the available commercial quota than any other State. Kangaroo populations continue to fluctuate, primarily in response to seasonal conditions. Populations are monitored annually by aerial surveys, fixed-wing and helicopter.

New South Wales NPWS issues licenses for the commercial harvesting of protected native plants from private property and some Crown lands such as state forests. The agency also issues licences for the commercial cultivation of protected plant species. However, some common species of native flora that are harvested commercially, particularly for the cut-flower/filler industry, are not protected species under current legislation and there is accordingly no statutory ability for the NPWS to manage or monitor the harvest or utilisation of those species.

HARVESTING UNDER COMMONWEALTH GOVERNMENT LEGISLATION

Better and more uniform data exist for some harvests that are undertaken in various Australian jurisdictions, but are listed under the Commonwealth's *Wildlife Protection (Regulation of Exports and Imports) Act 1982.*

Around 62% of the available quota for kangaroo harvesting takes place in Queensland. Nationally, around 63% of the total quota allowed is used. The quota levels set are well within the estimated population levels of the commercial harvest area. Numbers of the Wallaroo (*Macropus robust us*) have almost doubled since 1998, showing the fluctuations that can occur from year to year. Between 1998 and 1999, populations of all kangaroos increased, particularly for the Wallaroo and particularly in Queensland. In 1999, population estimates for South Australia, New South Wales and Western Australia were based on fixed wing aerial surveys. Population estimates for Queensland were based on extrapolation of helicopter counts in monitor blocks using line transect methodology.

In South Australia and New South Wales, however, the population of Red Kangaroo (*Macropus rufus*) declined over these years. For animals whose numbers fluctuate dramatically with rainfall, statistics from such a short period of time may have limited use.

Beginning in 1997, New South Wales has included a provision for kangaroos previously killed in the commercial zone under non-commercial permits to be included as an identified component of the quota. In 1996, South Australia refined the setting of the commercial harvest quota to provide for the separate identification of a sustainable use component to the quota and an additional land mitigation component. This latter component is to be released only when there is an identified threat to land management goals in areas where the 'sustainable' component of the quota has been taken.

Export data for Australian native plants and animals are available for permits issued. Over 250 species were subject to permit approvals, most for small and/or occasional quantities. There is substantial year-to-year variation in the volumes of individual species exported. Data are collected and reported in a range of units such as 'stems' and 'bunches' which makes comparisons between years and the determination of any trends difficult. It has not been possible to assess the significance of harvest levels of specific taxa, or the location of harvest.

Data on plants and animals exported illegally come from prosecutions and will thus be an underestimate. From the data provided it is usually impossible to determine what species are involved in prosecution or what crime was involved unless each individual file is analysed, a task beyond this report.

Management Plans for Sustainable Harvesting [BD Indicator 17.1]

Overall there are 11 management plans approved under the *Wildlife Protection (Regulation of Exports and Imports) Act 1982*. Most of the management plans are for species of kangaroo.

The Act provides for management programs to be declared where there is sufficient information available on the biology of the species proposed for harvesting to ensure that the activity will not be to the irreversible detriment of the species, or its habitat. Management programs are usually administered by state or territory government agencies and reflect state/territory-wide management for the particular species concerned.

Harvesting Indicators: An Overview [BD Indicators 8.1, 8.2 and 17.1]

Analysis of the available data for Indicators 8.1, 8.2 and 17.1 reveals that, with the exception of kangaroo harvest and permit data and flora harvest data from Western Australia, there is an acute lack of accessible comparable data for most states on harvesting of native flora and fauna.

There is a urgent need for improvements in the reporting system for prosecutions involving illegal exports of Australian wildlife so that it is possible to determine what species are being threatened.

Fisheries

Australia has sovereign rights to explore and exploit resources within the Exclusive Economic Zone (EEZ), which is the second biggest in the world, and also has a claim over the continental shelf where it extends beyond the EEZ outer boundary.

It also has the responsibility to manage and use these resources wisely, as well as to conserve biodiversity. The total Australian Marine Jurisdiction (AMJ), which includes the EEZ and continental shelf off the Australian mainland and external territories, covers a total surface area of around 16 million square kilometres - about twice the size of the Australian mainland cellspacing="2"

Australian fisheries are based on the harvest of wild, native organisms, and this represents a highly significant direct use of elements of the species level of biodiversity.

Traditionally, fisheries management has concerned itself primarily with the management of targeted fish stocks.

In recent years, fisheries management has broadened in scope to include consideration of the incidental mortality or injury of other species and to the effects of fishing operations on the environment. Such an approach is termed 'ecosystem management', a principle that dates from reform of national fisheries policy in the late 1980s to include ESD.

Proportion of Numbers Collected Over Size of Reproducing Population [BD Indicator 8.3]

Catch levels and management of major fisheries stocks are covered in the *Coasts and Oceans* Report.

Fisheries Bycatch [BD Indicator 8.4]

Bycatch species are those species that are not targeted in fisheries operations and are caught or affected incidentally. The definition of bycatch is complex, and can vary from fishery to fishery. Bycatch includes other fishery species, some of which may be retained and sold, as well as other species, including marine mammals and seabirds. In some fisheries, especially those using trawl techniques and comprising a mixture of species, bycatch is common due to the gear and methods used. In others, the specific nature of the gear and methods used mean that lower bycatch rates are possible. In general, bycatch describes a subset of the actual species present, with invertebrates and algae either grouped or ignored.

Published reviews of fisheries bycatch note that data are very poor for most fisheries, and that an accurate overview of the whole fisheries sector is not available. In some fisheries, discarded catch need not be recorded in fishers' logbooks and, where such recording is required, compliance may be low. This situation, in part, reflects the difficulty of identifying many species that are caught incidentally. In some fisheries, the bulk of bycatch is discarded, usually with high mortality rates. There is a significant bycatch problem in some fisheries, notably those utilising trawl and longline methods. The problem appears to be reduced in highly targeted fisheries such as those using jigging or diving.

OTHER EFFECTS OF FISHERIES ON BIODIVERSITY

'Ecological effects' refers to a range of other impacts of fisheries relevant to the species level of biodiversity. Recognition of such effects is a result of the beginning of the adoption of an 'ecosystem management' approach to fisheries, where management is intended to cater for impacts other than on the target stock. Ecological effects of some major Australian fisheries include:

- removal or mortality of one or more species (target or non-target), to the point where populations of that species may be rendered vulnerable
- the effect of 'fishing debris', such as the entanglement, pollution and possible drowning of wildlife in discarded or lost lines and nets
- impact on feeding or other behaviour of species such as Dugongs (e.g. through disturbance and boat-strike by vessels)
- disturbance of habitat, especially of seabed plant and animal communities through trawling
- indirect changes or destabilisation of ecosystems through the removal of important predator or prey species that may affect other populations
- impact on scavenging populations through the discarding of large quantities of bycatch and processing waste.

The effect of fisheries such as the Bass Strait Central Scallop Fishery on the biodiversity of benthic communities is likely to be high, but few reliable data are available.

Management Plans for Ecologically Sustainable Harvesting

Management plans are now provided for in all Australian fisheries legislation. Fisheries without management plans are managed in accordance with the general requirements of the relevant legislation. Although the situation across jurisdictions varies, the generally small proportion of fisheries with management plans, and the even fewer catering for bycatch, is a matter of concern from a biodiversity perspective.

Given that most of the legislation is less than one decade old, and that it has only been in recent years that ecosystem management and bycatch have been accorded real attention, the status may be viewed as transitional, and one that should be expected to improve markedly in the next few years and monitored accordingly. Requirements for environmental assessments and management plans under the EPBC Act and related legislation will speed up this process. A remaining issue even where management plans exist is of coordination of management where stocks cross jurisdictions.

The Policy Setting

Some policy and management changes have been initiated in the past decade affecting both target stock management, bycatch and other environmental effects. For example, Dugong Protection Areas have been introduced in waters adjacent to Queensland where the use of mesh nets are restricted or prohibited.

At the national level, a NBP was released in 1999 following cooperative development of the Commonwealth, states and territories, and expressly stated as being consistent with the National Strategy for Ecologically Sustainable Development and the NSCABD. The Policy states several guiding principles that should be adhered to in all legislation and management plans, including: promotion of stewardship, cooperation and transparency, integration of short-term and long-term approaches, and application of the precautionary principle. The overarching objective of the Policy is to ensure that bycatch species and populations are maintained at sustainable levels.

The NBP committed the Commonwealth to the development of bycatch action plans (BAPs) for all fisheries through a defined, multistep policy formulation. The Policy lists a range of strategies and a checklist of considerations to be taken into account when developing BAPs. BAPs do not have legal standing, and are not based on minimum standards of performance.

Effectiveness of Bycatch Controls [BD Indicator 17.2]

Investigation of Commonwealth fisheries (Barratt et al. 2001) suggest that 114 or 79% of Australia's 144 commercial fisheries have low to very low discard rates (defined as <25% of total catch discarded) while 11 fisheries have high discard rates (>50% discarded).

Fishing conducted by longline and dropline has increased in intensity, while the estimated level of bycatch for some significant fisheries is very high. The Northern Prawn Fishery, Southern Bluefin Fishery and South East Trawl Fishery, in particular, have high levels of bycatch and have, at least, a significant effect on marine biodiversity. For example, discard figures of 95, 83 and 50 to 86% are reported for these three fisheries, respectively. Some 30 000 to 60 000 t of marine life might be discarded to harvest 10 300 t of Northern Prawns. The 'discard' may involve over 500 species including turtles, snakes, sawfish, sharks and seabirds.

Various strategies are being investigated or adopted in other jurisdictions that could help reduce bycatch levels. In Queensland, the East Coast Trawl Fishery management plan includes the compulsory use of turtle excluder devices (TEDs). In New South Wales, research is being undertaken to establish methods of reducing juvenile fish bycatch in estuarine prawn and fish hauling. Release devices are also being introduced into prawn trawl fisheries in estuarine and ocean areas. In Western Australia, the issue is being attended to through action plans for individual fisheries. In general, the efficacy of the various bycatch reduction strategies and technologies requires ongoing monitoring.

ALTERED FIRE REGIMES

Fire Regimes

Fire is as much a part of the natural environment of Australia as wind, sun and rain. As a natural element, it has not only helped shape the environment, it also has been among the driving forces in the evolution of native fauna and flora (Gill et al. 1981; Whelan 1995). This has contributed to adaptations such as resprouting of many species after fire, especially the dominant eucalypts. The seeds of other plant species, such as hakeas and banksias, are held in the canopy and released by the heat of a fire. Smoke is an important cue for the germination of hundreds of plant species, particularly in south-west Australia (Roche et al. 1997). Whereas one fire alone may determine the response of an organ or organism, consideration of the fire regime (i.e. type, frequency, season and intensity of a fire experienced at a specified location, Gill 1975) is necessary to better understand responses of species and assemblages of species. A combination of fire and other environmental factors, such as life-history stage, plant condition, fire edge to area ratio, and the post-fire environment, may interact to affect biological patterns. Interactions with grazing animals and weeds can also affect the response of native biota.

Fires help create and modify the mosaic of landscape and biological patterns within their sphere of influence and may increase the susceptibility of some areas to erosion. The complexity of fires and the way they may change native ecosystems should not be underestimated; fire effects can still be detected 20 to 30 years after the event. Much is still to be learnt about the long-term effects of both single and repeated fires on natural ecosystems.

Despite the intimate relationship between fire regimes and the biota of Australia, fire is perhaps one of the least understood of the 'natural elements'. The popular view is that it is an element of destruction. Images on television of raging bushfires both in Australia and other countries tend to reinforce that perception, particularly within the ever-increasing urban community. Such coverage portrays fires as individual events, whereas it is the conditions of the ecosystems and their characteristics, together with the history of fires and their properties that determines the effect of fires (Bradstock et al. 2001).

Fire and Humans

The use of fire by humans has made an indelible impact on many species, including humans. For millennia, Indigenous peoples have used fire for a range of purposes including warmth, hunting, communication, ceremonies and cooking (Bowman 1998). Prior to the European occupation of Australia, landscape burning was widely used by Indigenous peoples and their effect on the environment is considered one of the most complex and contentious issues in Australian ecology (Bowman 1998). Indigenous peoples' knowledge of landscapes and use of fire remains extraordinarily detailed, particularly in central and northern Australia. A recent study on fire patterns and their impacts across northern Australia illustrates the lessons that can be learnt from traditional burning. It also emphasises the different fire regimes that can occur on different land tenures, making the description of fire patterns across the country even more complex.

A unique study in the remote desert west of Lake Mackay (WA) (N. Burrows, unpublished data) provides quantitative evidence of traditional patterns of burning in arid Australia. Aerial photographs taken in the early 1950s, when Pintupi people still lived a traditional lifestyle, reveals a broad-scale pattern of numerous small patches, 75% of which were less than 32 ha, and occasional large fires up to 6000 ha. In contrast, images taken in 1988 reveal a very different pattern of burning. The main exodus of Pintupi people from the remote desert into European settlements occurred in the early 1960s (N. Burrows, pers. comm.) and the cessation of traditional burning in the region is considered to have led to the much larger fires. Information such as this can be used to support contemporary fire management, especially where the reintroduction of more traditional approaches is the goal.

In temperate Australia, the ecological connectivity and functionality has been disrupted by broad-scale clearing and other major modifications, and traditional Indigenous land management has long been replaced by management for very different aims (e.g. Williams & Gill 1995). However, a range of other techniques can be used in southern Australia to gain some understanding of historic fire regimes and their effect on the biota. These include palaeoecology, dendrochronology, ethnography and an understanding of the life-history strategies of plants and animals (Williams & Gill 1995; Kohen 1996; Benson & Redpath 1997).

PRE-CONTACT INDIGENOUS, AND CONTEMPORARY FIRE REGIMES OF THE SAVANNA LANDSCAPES OF NORTHERN AUSTRALIA

Ethnographic, historical and contemporary observations concerning traditional burning, while sparse and geographically biased towards coastal and subcoastal regions in northern Australia, consistently show that burning was undertaken throughout the dry season following landscape patterns of the curing of grassy fuels, but particularly in the early-season to mid-dry season under cooler, milder conditions for fire. Burning of clan estates was/is undertaken systematically and purposefully. Contemporary evidence (from coastal and subcoastal situations) indicates at least half of any clan estate might be burnt in any one season. It is evident that, in accord with regional human population densities, burning was undertaken more frequently in higher rainfall coastal and subcoastal regions.

Based on regional mapping of fires from satellite imagery (mostly National Oceanic and Atmospheric Administration-Advanced Very High Resolution Radiometer, NOAA-AVHRR and LANDSAT) from the 1980s, two broad contemporary patterns have been identified concerning the application of fire in northern Australia. In north-west and northern Australia, and around the Gulf of Carpentaria, vast tracts are burnt annually, typically by intense wildfires late in the dry season. Conversely, elsewhere across northern Australia, but especially on more productive pastoral lands, landscape burning is infrequently applied.

Major differences (and similarities) between traditional Indigenous peoples and contemporary fire regimes may be summarised as follows:

- Whereas burning was undertaken across northern Australia under Indigenous custodianship, burning today is concentrated mainly in non-pastoral, relatively high rainfall regions, especially in the Kimberley, in the Top End, and around the Gulf of Carpentaria.
- Whereas burning traditionally was concentrated in the early- to mid-dry season, today it generally occurs mostly in the late dry season. Generalisation of the contemporary situation, however, masks considerable early dry season burning in some limited locations (e.g. Darwin region, Kakadu and Litchfield National Parks).

- Whereas it is evident that burning traditionally was/is undertaken systematically for a diverse range of purposes, today where burning occurs, it often emanates from uncontrolled wildfire.
- Importantly for biodiversity conservation, whereas an essential feature of burning by Indigenous peoples was/is that it tended to be highly patchy and thus contributed to developing habitat heterogeneity, today northern Australian savanna landscapes are either burnt frequently by typically intense, extensive fires, or seldom burnt.

Current Fire Regimes

In southern Australia, one of the main objectives in the use of prescribed burning is to reduce the rates of spreads and intensities of 'wildfires'. Experience, rather than empirical data, seems to be a major factor in assessing the value of fuel-reduction burning as a management technique. Altered fire regimes have been implicated in local extinctions of several vascular plant species across Australia (Gill & Bradstock 1995; Keith 1996) and inappropriate fire regimes have been associated with 19 plant species threatened with extinction at the state or Commonwealth level (Leigh & Briggs 1992). Gill and Bradstock (1995) also list 19 examples of local plant extinctions that span a wide range of life histories, habitats and locations. The changes in habitat structure that come with the decline and elimination of woody plant species under frequent fire regimes have demonstrable implications for the persistence of other groups of biota (e.g. Catling 1991; York 1999).

Single fire events, especially of high intensity, can eliminate species as recorded for alpine conifers in Tasmania (Kirkpatrick & Dickinson 1984). Other kinds of fire regimes have also been implicated in declines and extinctions of plants. For example, the decline of woody heathland understorey species under low frequency fires has been linked to competitive exclusion (Keith & Bradstock 1994) and the lack of fire in isolated remnants in the highly cleared landscapes of southern Australia (Lunt & Morgan 2001) is of concern.

Many Australian birds have declined since European settlement. The involvement of altered fire patterns in this decline has long been recognised (Ashby 1924). The 2000 Action Plan for Australian Birds estimated that changes in fire management affects 43% of mainland bird species (Garnett & Crowley 2000). In many cases, birds that are threatened by altered fire regimes require long-unburnt vegetation and intervals between fire longer than those which have been imposed since European settlement (Woinarski 1999). This is especially so for birds of heathlands, mallee and coastal 'scrub', and birds reliant on hollows for nesting or roosting. A major change in fire regimes is not needed to trigger biodiversity loss: even minor changes in fire regimes may be critical for some bird species and can lead to almost imperceptibly gradual, but inexorable decline, especially where habitats have been extensively fragmented (Brooker & Brooker 1994).

The study by N. Burrows et al. (unpublished data) in the Western Desert shows that since the 1950s fire patterns in the hummock grasslands of the Great Sandy Desert have changed from small interlocking burnt patches to a simpler mosaic consisting of large tracts of long unburnt or recently burnt vegetation. This pattern has been repeated across arid and semi-arid Australia (i.e. Griffin et al. 1983; Griffin & Friedel 1985) and such changes have been implicated in the decline of small mammals in these areas (Burbidge & McKenzie 1989). Other factors such as increased levels of predation, drought and competition for resources from exotic animals also appear to have been important in the demise of small mammals (Morton 1994).

Monitoring fire patterns [BD Indicator 6]

Remote sensing by satellites has opened the way for mapping and monitoring the areal extent of fire activity on a continental scale. The NOAA-AVHRR satellite sensor has already been used for several years for real-time fire monitoring by the WA Department of Land Administration to identify potentially damaging bushfires in northern Western Australia and the Northern Territory. This technique was also used for a pilot study to assess fire patterns across Australia between April 1998 and March 2000 (Craig et al. 2000).

LEGISLATIVE, POLICY AND MANAGEMENT FRAMEWORK FOR FIRE MANAGEMENT

Historically, much of the legislation concerning fire management in Australia has been about fire prevention and suppression, aimed at minimisation of loss to property and life. Legislators respond to deaths in bushfires by attempting to impose control on fire. In many cases, this legislation narrows the range of acceptable or achievable fire regimes, and in some cases outlaws regimes which are required for the conservation of some biota (Hughes 1995).

At the state level, there are some strong legislative responsibilities related to fire planning and management for biodiversity conservation. For example, the fire management plan for Tarawi Nature Reserve in western New South Wales operates under the *Rural Fires Act 1997* (NSW) which defines the statutory obligations of the land manager and provides for establishment of District Bushfire Management Committees as a means of integrating fire management across landscapes that comprise multiple managers with varying goals.

The plan also operates under the *National Parks and Wildlife Act 1974* (NSW), which defines the role of Nature Reserves and requires that fire management is not in conflict with the Plan of Management adopted for the reserve, and the *Threatened Species Conservation Act 1995* (NSW), which defines requirements for impact assessment, planning and implementation of recovery for listed species.

Keith et al. (2001) proposed several principles and approaches for the use of fire as a management tool that recognises the importance of setting explicit goals, precautionary management practices, experimentation, risk assessment and the need to implement monitoring so that management practices can be evaluated and potentially modified. An important element of this approach is to try and include the unpredictable (especially unplanned fires) in management planning.

Reducing the impact of altered fire regimes [BD Indicator 21]

In recent years, there has been an increasing emphasis on the development of fire management plans that specifically incorporate biodiversity conservation. The current regimes for conservation reserves in each state are described as well as national parks that the Commonwealth government has responsibility for. It does not, however, address other land tenures such as State Forests and Indigenous peoples' lands (except for jointly managed conservation reserves) where fire is also important in the maintenance of biodiversity.

New South Wales

The New South Wales NPWS has responsibility for 4.6 million hectares of national parks and reserves. The Service has a systematic program in place to develop fire management plans that specifically address biodiversity conservation for over 200 fire-prone reserves. The system uses a flexible, adaptive management approach based on fire regime thresholds defined for broad vegetation groupings. This enable managers to adjust fire management strategies on an ongoing basis to encompass changes in fire regimes that result form both planned and unplanned ignitions.

Effective use of the system requires commitment to continual monitoring and mapping of fires and assessment of their additive effects (the fire regime). Strategies for the protection of life and property and the management of heritage sites are also addressed. The resolution of these objectives with biodiversity conservation is accomplished through a zoning system. The use of Geographic Information Systems (GIS) is an integral element, and a range of other databases are being developed to support the planning process. More than 30 Reserve Fire Management plans have been finalised, and over 150 are in draft form or advanced preparation.

At a broader scale, the first round of Bushfire Risk Management Plans have been completed for major bushfire districts in the state. This initiative of the Rural Fire Service is a broad-scale attempt to coordinate cross-tenure planning for protection of lives and property, while taking into account the requirements of specific land tenures (forests, farms, parks and private property). A major focus of the Plans has been to ensure that bushfire management is ecologically sustainable. A process for assessment of risk to humans along with the ecological requirements of threatened species and biological communities is a feature of the planning approach. The completion of Risk Management Plans has necessitated major inputs from state government agencies (e.g. State Forests, Land and Water Conservation, NPWS), local government and community groups.

RISK ASSESSMENT AND MANAGEMENT APPROACHES TO BIODIVERSITY

An increasingly important issue is the applicability of risk management approaches to biodiversity (e.g. AS/NZS 4360, revised version, Standards Australia 1999) and the Standards Australia (2000) handbook. This Standard is based on subjective interpretation of the likelihoods and consequences of hazards, and is becoming the standard for the development of risk management systems in industry and institutional risk management in Australia.

Conservation biology embarked on a different path for risk assessment two decades before the Australian standards were developed. From its earliest inception, risk assessment for biodiversity has typically been informed by model-based (Shaffer 1981; Boyce 1992; Burgman et al. 1993; Possingham et al. 1993), or at least rule-based (IUCN 1994) risk analysis. These methods have continued to develop into a formidable array of analytic and decision-support tools. These risk assessment tools are valuable mainly because they set the decision-making process in a formal framework in which the costs and benefits of management alternatives can be explored and updated, they are relatively transparent and free of semantic ambiguity, and they provide some assurance of internal consistency. These benefits are difficult, if not impossible, to achieve in a subjective risk assessment framework.

An example of the application of subjective risk assessment methods is provided by the protocols used by the Interim Office of the Gene Technology Regulator, and its Genetic Manipulation Advisory Committee (GMAC). GMAC relies on subjective interpretation of available data to evaluate the risks and consequences of the various potential ecological, social and economic costs and benefits posed by GMOs. The lack of transparency and robust interpretation of data has lead the Commonwealth government to invest several million dollars in a CSIRO study aimed at developing formal, quantitative tools for assessing the ecological risks posed by genetically modified species.

Western Australia

Fire management on public land in Western Australia is primarily the responsibility of the Department of Conservation and Land Management (CALM). For each area managed by CALM, a Management Plan is prepared and approved. Each plan contains a section on fire management that explicitly deals with the need to manage for biodiversity through the application of appropriate fire regimes or fire exclusion. Plans are developed both at a regional level for groups of similar geographically related areas, as well as for single conservation parks and nature reserves.

There are 39 approved Management Plans for terrestrial conservation reserves in Western Australia (includes National Parks, Conservation Parks and Nature Reserves) and 26 Management Plans in preparation, out of a total of 1283 conservation reserves. For many of the conservation areas that do not yet have a fully developed management plan, interim management guidelines (IMG) are prepared to enable managers to proceed with operations considered necessary for the conservation and protection of the biodiversity and ecological values of the reserve. Even though these IMG are primarily focused on fire protection strategies, they also include consideration of fire to enhance the ecological values of the area.

Management plans for forested areas provide for a variety of fire regimes. One of the key management objectives for fire is to minimise the likelihood of entire reserves being burnt at the one time. Plans include burning for fuel reduction on varying cycles, vegetation management (with a medium-term to long-term rotation) and reference areas with no planned burning. Plans for fauna-rich areas have long-term fire regimes. Monitoring programs are also implemented in some areas to assess the effect of the prescribed burning regime on indicator flora and fauna (e.g. vulnerable species).

In south-west Western Australia, CALM has undertaken many years of research into fire ecology and is using this information to apply varied fire regimes that will conserve natural systems as well as protect life and property. An interesting research tool has been the determination of the past fire history of jarrah forests from fire scars on the Grasstree (*Xanthorrhoea* spp.). Grasstrees, which live over 200 years, provide an insight into the fire frequency prior to European settlement (CALM 2000b). These data, combined with information from historical literature and consultation with the Nyungar Indigenous community from the south-west area, will better inform current fire management practices for the conservation of natural systems.

Australian Capital Territory

The Nature Conservation Strategy of the Australian Capital Territory (from the *Nature Conservation Act*) identifies fire as a key threatening process to biodiversity. This is recognised in the ACT Bushfire Fuel Management Plan 1998 (BFMP), which identified 'the maintenance of biodiversity and natural processes' as a management objective. The BFMP is a requirement of the *Bushfire Act 1936* (ACT) and provides a framework for fire management on most government lands. The plan provided some coarse-scale recommendations for the conservation of biodiversity, based on general principles of fire ecology. Recommendations include indicative burning intervals for various forest and woodland communities. However, extensive research and monitoring is required to develop appropriate burning regimes for wildlife habitat management and for the conservation of biodiversity.

ACT Parks and Conservation is the agency with primary responsible for fire management in conservation reserves, and is required to develop fuel management plans for these areas. It is the role of the Bushfire Service to implement these plans, and they have the power to accept or reject them.

South Australia

There is no formal process for fire management in South Australia. In the *Native Vegetation Management Act 1985*, burning was considered as vegetation clearance and needs approval by the Native Vegetation Council. There are no explicit criteria for the approval of burning, and instead the Council assesses each application in terms of the NVM Act and also considers advice from scientific officers. Fire management in conservation reserves is considered mainly from the perspective of fire risk and prevention. Although work is in progress by National Parks and Wildlife to develop a decision support system for ecological burning, there are no policy or management plans explicitly directed at fire management for the enhancement and conservation of biodiversity.

Northern Territory

The Bushfires Council (BFC) of the Northern Territory is a statutory body set up to coordinate fire management within the Northern Territory and to provide an organisational framework. Established under the *Bushfires Act*, the BFC operates under policy guidelines designed to achieve fire management objectives. Although the primary objective of the BFC is to reduce the total area burnt by wildfire in the Northern Territory, the maintenance of native ecosystems, by the use of appropriate fire regimes, is a key purpose of the BFC fire management strategy. To enhance the coordination of fire management, the Northern Territory has been divided into nine fire control regions, based on characteristics such as land systems, vegetation type, climate, location and service centres.

Although a fire management plan is written for each region, the focus is primarily on operations for prevention and mitigation of wildfires, rather than detailing appropriate fire regimes for the conservation of biodiversity.

As an umbrella organisation, the BFC works closely with the PWC to manage fire within Conservation Parks and Reserves across the Territory. The PWC is responsible for about 90 parks and most of these have plans of management. Each plan has a lifetime of five to 10 years and many plans are now in their second version. The PWC recognises that fire plays an important role in reaching their objective of maintaining and encouraging optimum biodiversity. In this light, most park management plans identify fire as an issue for biodiversity. Several parks also have Fire Management Plans, but specific guidelines outlining appropriate fire regimes for the long-term conservation of biodiversity are generally not provided.

Monitoring programs have been established in both Litchfield and Nitmiluk National Parks to increase understanding of the effects of fire on the vegetation communities, and to better inform fire management practices.

Tasmania

In Tasmania, the Department of Primary Industries, Water and the Environment (DPIWE) is responsible for the development of fire management plans. Generally, management plans are prepared only for parks where fire protection is an issue. Prior to 1996, management plans were almost totally concerned with excluding fires from areas. Over the past four years, ecological burning has increased in importance in the plans, but most plans still focus on reducing fuel loads. Because of the wet and cool climate and associated vegetation types, fires are not considered an issue in many of Tasmania's parks. Consequently, management plans for fire are not developed for all areas.

There are four implemented fire management plans for conservation parks and reserves in Tasmania that specifically consider the interactions between fire and biodiversity. There are a further ten plans for which fire-biodiversity interactions are considered, but are not a significant component. Four fire management plans do not consider biodiversity, but these are under revision and are not being implemented.

The most extensive fire management plan considering biodiversity conservation in Tasmania is for the south-west Tasmanian lowland buttongrass moorlands, in the southern half of Tasmania's Wilderness World Heritage Area. This plan provides burning prescriptions, based on detailed modelling, to enhance the development of a high level of species richness and structural diversity. The plan is intended to be an interim one until a more comprehensive plan is produced during the next three years, and incorporates principles of adaptive management to ensure research is undertaken so that management prescriptions can be refined if necessary.

Queensland

The Queensland Parks and Wildlife Service (QPWS) drafted a state-wide fire policy in mid 2000 specifying that fire management in the conservation estate will be for ecological purposes. Fire management plans are drawn up by QPWS for individual protected areas. Most fire management plans consider biodiversity; however, the detail to which they address biodiversity-related issues varies. A new template structure was adopted in Queensland in 2000, which will require that biodiversity aspects be detailed in all fire plans over the next two to three years. The protection of ecological systems and hence, biodiversity, is one of the two main purposes for these fire management plans, the other being the safeguarding of life and property.

Victoria

Fire management on Victoria's public land is the responsibility of the Department of Natural Resources and the Environment (DNRE). The DNRE has been moving gradually from a sole focus on fire protection to one of fire management that takes into account the ecological effects of fire. All public land within Victoria is covered by five regional fire protection plans that have been established under the Code of Practice for Fire Management on Public Land (DCNR 1995). Determined in consultation with relevant park and forest managers, these plans divide the State into five fuel management zones, of which two zones incorporate management for the protection of the flora and fauna values of the region. Victoria is also starting to accumulate fire management plans for specific parks and conservation reserves. Although management plans for national parks outline objectives and strategies for fire management within the park, specific guidelines for ecological burning are generally lacking.

The paucity of ecological data relating to fire regimes has meant that consideration of biodiversity in fire management has been minimal. However, recent initiatives by DNRE and Parks Victoria have begun to address this issue. Workshops with managers of public land were held in Victoria during 1998 to discuss fire management for the conservation of biodiversity. A set of guidelines for ecological burning has been drafted, and pilot studies established to research and develop ecologically based fire regimes for a variety of vegetation types. The ultimate aim of fire management in Victoria will be to provide sufficient information on each Ecological Vegetation Class across the state, for informed decision making about fire frequency and requirements.

Commonwealth

The Commonwealth, through Environment Australia, manages parks and reserves established in those parts of Australia which come under its direct responsibility, such as the External Territories. The EPBC Act is the principal Commonwealth legislation for establishing and managing protected areas. The Director of National Parks is a statutory office established under the Act with responsibility to administer Commonwealth reserves. Six national parks and five national nature reserves are declared under the Act. The other nine reserves are either marine parks or botanic gardens and will not be considered further here. Three of the six Commonwealth national parks, namely Kakadu and Uluru-KataTjuta National Parks in the Northern Territory and, more recently, Booderee National Park in the Jervis Bay Territory are managed jointly by the Indigenous owners and Parks Australia. The other three national parks protect unique island ecosystems in the Territory of Cocos (Keeling) Islands, the Christmas Island Territory (in the Indian Ocean) and the Norfolk Island Territory (in the South Pacific). The management of exotic plant and animal species, rather than fire, is a key management issue on these islands.

Kakadu National Park: The Kakadu National Park Plan of Management (Kakadu Board of Management and Parks Australia 1998) has four main aims in relation to fire management. These are to:

- promote traditional Indigenous peoples' ways of burning within the park
- protect life and property within and adjacent to the park
- restrict fire from spreading so that it does not enter or leave the park
- maintain biodiversity through effective fire management of species and habitats.

Langton (2000) stated that the continuance of traditional Indigenous fire management in parallel with park management efforts in Kakadu is particularly noteworthy because of the high standard of research and documentation on these efforts and the ready acceptance of the importance of traditional knowledge of fire.

Uluru-Kata Tjuta National Park: The most recent Plan of Management for Uluru-KataTjuta National Park (Uluru-KataTjuta Board of Management and Parks Australia 2000) has the following aims for fire management:

- maintain traditional Anangu burning practices and promote their integration into scientific knowledge, to protect and enhance the Park's biodiversity

- protect life, property and culturally significant sites and mitigate the effects of wildfire
- maintain community education and interpretation programs dealing with the role of fire in the Park
- maintain a research and monitoring program and operate within a regional context, and to help neighbours suppress wildfires when resources are available.

Tjukurpa ('the Law' governing that Indigenous community's actions and culture) is a guiding principal for fire management in the Park. A patch burning strategy is used, based on traditional patterns of burning. Anangu (the traditional owners) and Parks Australia work together to determine which areas should be burnt each year. This combines traditional ecological knowledge with the use of GIS (Allan 1997) and the results of ecological studies by western scientists. Saxon (1984) has also been used to guide fire management in the Park but it is acknowledged that this requires updating. Comprehensive, long-term monitoring is required to determine the effects of the current approach to fire management on biodiversity in the Park.

Human-induced Climate Change

The Third Assessment Report from the Intergovernmental Panel on Climate Change (IPCC) was due to be released in mid 2001, and the Summaries for Policy Makers for the three working groups were released in early 2001 (IPCC 2001). This report and the regional climate projections released by the CSIRO in May indicate that Australia can expect to be generally warmer and drier, but with increased floods and storm surges. Australia's natural systems will have difficulty adapting, with vulnerable areas including the Great Barrier Reef, alpine ecosystems, wetlands and riverine systems and woodlands.

Climate change will affect biodiversity and it presents serious challenges for management aimed at conserving biodiversity. Several potential changes arising as a result of climate change will directly and indirectly affect biota. The degree of adverse climatic effects depends on the ability of the system to adapt to climate change.

The difficulties in predicting the effects of climate change arise because:

- detailed regional forecasts of potential changes in climate have only emerged recently
- understanding of the potential response of biota and ecological systems to these changes is limited.

Regional forecasts of climate change [BD Indicator 7]

The Atmosphere Report provides a detailed discussion of forecasts of climate change under an enhanced greenhouse effect. By 2030, most of Australia will be warmer by 0.4 to 2.0C. For 2070 the warming is 1.0 to 6.0C (estimates for 2030 and 2070 are subject to spatial variation). In summer and autumn, projected rainfalls for most of Australia are-10% to +10% by 2030 and -35% to +35% by 2070 and tend towards an increase. In winter and spring, most locations tend towards decreased rainfall and are estimated at-10% to +5% by 2030 and-35% to +10% by 2070 (CSIRO 2001). Soil moisture changes are expected as a result of changes in rainfall characteristics and evaporation. Higher average temperatures are likely to increase evaporation. The global increase in sea level is expected to be between 9 and 88 cm by 2100, or 0.8 to 8.0 cm per decade.

Direct Effects of Climate Change

The potential responses of biodiversity to climate change can be considered in terms of changes in the distribution and abundance of taxa, species performance, and interactions between species which have implications for the structure and function of ecosystems (IPCC 1998). The potential ecological trajectories induced by climatic change encompass processes over the entire range of magnitudes from the level of leaf physiology to biome physiognomy and distribution. At least two conceptual views of biome response to climatic change have been proposed: ecotones gradually shifting in space, or ecosystems rapidly undergoing change over large areas in response to catastrophic disturbance such as drought (IPCC 1998). The view of gradually shifting ecotones in space arises largely from an emphasis on demographic processes, whereas the concept of catastrophic change arises largely from an emphasis on ecosystem function (water and nutrient) processes. These two models present different images of future change and biosphere responses.

CHANGES IN THE DISTRIBUTION AND ABUNDANCE OF TAXA

Large-scale changes in the distribution of species and biomes has occurred. However, the anticipated changes in global climate are expected to occur at a rate most biologists acknowledge as simply too fast for evolutionary processes, such as natural selection, to keep pace. Such constraints on the ability of species to adapt to their rapidly changing habitat could substantially increase their probability of extinction. In addition, landscape fragmentation related to human activities will markedly limit the opportunity for some species to migrate. It has been suggested that habitat destruction and climate change will act together, setting the stage for greater rates of extinction than when considering human encroachment alone (IPCC 1998).

Species Performance

The intrinsic ability of a species to colonise will depend on its ecological characteristics, including reproductive rate, viability and growth, the way it disperses, and its ability to tolerate inbreeding. Species disperse at different rates, which may result in dramatic alterations of the species composition of all biological communities. The biology of a species will be crucial in determining the rate at which it can respond to climate change. A species can extend its geographical range only if humans move it, or by natural processes individuals disperse to, and establish in, areas beyond their current distribution.

Physical constraints of a locality will restrict migration and increase vulnerability. For species on small offshore islands, southernmost coasts, the northern boundary of the central Australian desert, the tops of mountains, riparian zones, and forest remnants separated by urban development, there may be no migration options. Soil differences, inadequate rainfall, or excessive wind may also prove to be barriers to migration (IPCC 1998). Conversely, changes in temperature and rainfall may remove an existing barrier to migration, such as frost. The physiological adaptations of most species to climate are conservative, and it is unlikely that most species could evolve significantly in the time allotted by the coming warming trend. Most of the available data on the effects of elevated carbon dioxide levels on vegetation have been derived from short-term treatments in controlled environments. Many of these studies report increased growth rates (particularly of C_3species), although there have been striking interspecific variations in responses to higher temperatures and carbon dioxide levels (IPCC 1998).

Response of Plant Communities

The ability of plant communities to accommodate climate change will be influenced by plant-soil-soil moisture interactions (IPCC 1998). This relationship among soil, soil moisture, vegetation and climate needs to be better understood in order to project responses.

Disturbance Regimes

Disturbance regimes are often a main factor in determining the suitability of habitat for species and, hence, will be of major importance in facilitating turnover from one species or vegetation type to another in response to climate change. Studies in other countries have identified the importance of altered fire regimes under climate change for a range of ecosystems. Fires are also integral in the dynamics of most Australian ecosystems. Therefore, the response to disturbance regimes must be considered when predicting ecosystem dynamics. Changes in climate and fuel dynamics will affect future fire regimes. Factors such as ignition sources will also be important.

Ecosystem Functions

In natural ecosystems, the timing of plant fruiting and flowering, which at least in the tropics is largely determined by the temporal distribution of droughts and rainy periods, may be adversely affected if rainfall patterns change. Most research efforts into the affect of elevated carbon dioxide levels do not include an examination of the plant's reproductive responses, despite their importance to ecosystem function.

As well as the effect on plants, pollinators may be affected by climate change. This can be expected to have significant consequences for plant reproduction. Long-lived species such as established trees might show a muted response to climate change, exhibiting substantial time lags assuming they are not significantly altered by other human activities or by catastrophic disturbance. Disturbances such as fire, however, will create opportunities for more rapid change by reducing the inertia of, at least, established forests. Even if adults of a species can tolerate changes in climate, their ability to produce propagules and the ability of those propagules to recruit to maturity may be adversely affected. Many Australian animals, particularly birds such as the Yellow-tailed Black Cockatoo (*Calyptorhyncus funereus*), are relatively long-lived. Without knowledge of the fecundity of these species, the presence of adults in a population may not be a reliable guide to the longer term persistence of a population or the species (IPCC 1998).

Parasitic and Invasive Organisms

The geographical distribution of many parasitic species are limited by the distributions of potential host species or by environmental constraints on the parasite's rates of development (IPCC 1998). The effect of changing climate will depend, therefore, to some extent on the response of the host to the altered conditions. Where members of the parasite community are important in mediating competition between hosts, this may lead to further changes in the structure of the host community and the possible extinction of particularly susceptible hosts.

If the temperature increases significantly, parasites and disease will do well as they are by definition organisms that colonise and exploit, particularly in relation to stressed individuals on the edge of their environmental range. The details of where and when these changes will occur, and what effect they may have on the distribution and abundance of species, are relatively poorly understood (IPCC 1998). Predictive models have focused principally on species important from an economic perspective, such as the cattle tick.

Human-induced changes to climate and habitat could dramatically increase the frequency of invasions by organisms from outside their current biogeographical boundaries. Therefore, planning for the movement and invasion of species into new areas is essential. However, understanding the consequences of global change on species invasions requires a much better understanding of the community and ecosystem roles of individual species.

Aquatic Systems

Changes in hydrology associated with climate change may have serious implications for wetland biota. Although some aquatic plants are able to survive water fluctuations, it is unknown how wetland biota will respond to the potentially major changes in rainfall and subsequent run-on and run-off characteristics. Several migratory birds that utilise Australian wetlands may also be affected if wetland dynamics alter under climate change. Potential changes to wetlands resulting from climate change will be aggravated by the modification of wetlands through draining, fragmentation and rising water tables (IPCC 1998).

The responses to climate change of the large, arid, ephemeral lake systems of interior Australia are difficult to predict. Although these systems already experience significant seasonal and interannual variations, and the associated ecosystems are attuned to this high variability, their resilience to long-term change in the frequency and intensity of events is less certain. Significant water level changes may occur for non-ephemeral lakes in dry evaporative drainages or small basins where, at present, evaporation is comparable with rainfall inputs. Estuaries and coastal wetlands have survived historical rises in sea level, usually by migration landward; salt marshes and mangroves have survived where the rate of sedimentation approximates the rate of local sea level rise; beaches have grown or decayed according to changes in prevailing winds and seas; and coral reefs have demonstrated the capacity to grow vertically in response to past rises in sea level. However, these past rates of adaptation may be insufficient for the higher rates of future rises in sea level, and in many cases landward migration will be blocked by human infrastructure, such as causeways, flood protection levees, and urban development, leading to a reduction in the area of the delta or mangrove (IPCC 1998).

Coral Reefs

Coral reefs and atolls in the region, and in neighbouring South Pacific countries, are among the most sensitive environments to rises in sea level and climate change, through potential inundation, flooding, erosion, saline intrusion and death of corals. Coral bleaching and decline from prolonged increases in seawater temperature can inhibit their capacity to grow at the rates required by sea level rise. Managing reef ecosystems such as the Great Barrier Reef may be more problematic as a result of climate change. Increased sea temperatures seem to be the main concern, rather than sea level rise per se. The potential sensitivity of coral reef ecosystems to climate change was demonstrated in 1998 when a global episode of coral bleaching showed that many reef corals live near the limits of their thermal tolerance. Where corals bleached, sea temperatures were several degrees above normal summer values and were some of the highest on record. Climate change is believed to be a major threat to coral reefs and could be a major driver of change across one of the most productive ecosystems on earth (AIMS 2000).

Minimising the effects of climate change [BD Indicator 22]

The international community is engaging in considerable scientific research to deepen understandings of climate change.

Australian scientists are prominent in many areas, with the focus on developing models of regional climates and carbon sequestration (through the CRC for Greenhouse Accounting). Research on the direct effects of climate change on biodiversity is relatively limited, with some work underway on the response of different organisms to elevated carbon dioxide levels.

SoE (1996) presented modelling results that illustrated changes in the potential distribution of several species under different climate change scenarios (Dexter et al. 1995). More recently, modelling has demonstrated similar effects on different species. These studies show the potential for significant effects on native species under the current models for climate change.

STABILISING GREENHOUSE GAS EMISSIONS

Much of the focus on minimising the effect of climate change has attempted to stabilise greenhouse gas emissions. Because of the global threats imposed by climate change, these approaches have also been undertaken at an international level. Australia became a party to the United Nations Framework Convention on Climate Change (UNFCCC) in 1992. The Convention aims to stabilise emissions of greenhouse gases at a level that would prevent dangerous human-induced interference with the climate system.

The UNFCCC has led to the Kyoto Protocol, agreed in December 1997. The Protocol aims, inter alia, to enhance the energy efficiency of national economies, protect and enhance sinks and reservoirs of greenhouse gases, promote sustainable forms of agriculture, limit and/or reduce emissions of greenhouse gases, and remove economic instruments that undermine such outcomes. The Protocol outlined emission targets for developed countries, with Australia having a target of restricting greenhouse gas emissions for 2008 to 2012 to an increase of 8% over 1990 levels. To date, Australia has submitted two reports (Commonwealth of Australia 1994, 1997) on its commitments under the Convention and these reports were reviewed by independent committees appointed by the UNFCCC Secretariat.

Carbon Sinks

Some mitigation actions that are promoted by Australia to meet greenhouse gas emission targets could have considerable effects on biodiversity, especially those related to carbon sinks. For example, the potential for vegetation clearance to be reduced to maintain carbon sinks could have a positive impact. Clearing of native vegetation is a major source of emissions in Australia and accounts for 13 to 18% of the total annual emissions.

Significant reduction in, or elimination of vegetation clearing, would remove of one of Australia's most serious threats to biodiversity, and would help meet international greenhouse gas commitments. Similarly, biodiversity will benefit both directly and indirectly from measures to improve national energy efficiency and put agriculture on a sustainable footing.

In calculating net emissions of greenhouse gases for 2008 to 2012, Australia is likely to be allowed to count the removal of carbon dioxide from the atmosphere to sinks such as vegetation and soils. The sequestration (or storage) of carbon into vegetation and soil sinks is one way to reduce net emissions of greenhouse gases. These sinks have the potential to be linked in with carbon 'credit' schemes that provide a basis to make carbon a tradeable commodity. Under such schemes, vegetation has a quantifiable value as a carbon store, irrespective of its value for biodiversity, and may provide stronger investment incentives for landscape revegetation.

Carbon sequestration schemes have focused mainly on plantation trees in Australia, although other opportunities exist to sequester carbon by adopting changes in land use. One example is the retirement of land from conventional agriculture, and promoting the thickening of understorey vegetation of native grasses and woody shrubs, which in turn become carbon sinks. These types of activities and strategies may benefit biodiversity if biodiversity conservation is treated as an integral or strong complementary aim. Revegetation strategies can markedly change the biophysical and ecological processes of a region and help reverse habitat degradation (e.g.

lowering water tables and the affects of salinity). However, poorly conceived strategies of this nature may be as bad for biodiversity as many current threats and human activities.

Despite the benefits for carbon sequestration and climate change, the planting of large areas with tree monocultures can have serious socioeconomic, as well as biodiversity, implications (e.g. Gill & Williams 1996). In many regional areas, the rapid growth of the plantation industry is changing employment opportunities and investment strategies and requires the upgrading of road infrastructure and networks for use by heavy transport.

Overall, the future of much of the biodiversity of the Australia is threatened by climate change. Australian governments must adopt enhanced measures to achieve genuine and significant reductions in greenhouse gas emissions and prevent human activities that seriously compound the potential effects of climate change on biodiversity.

Pollution

Impacts of pollution on biodiversity [BD Indicators 5 and 20]

The release of pollutants into the environment can kill organisms outright, change the biogeochemical conditions and processes occurring within a system and result in systemic changes that degrade habitats and make ecological processes dysfunctional. Biodiversity associated with sites intensively used by humans may be most at risk, although the non-point based effects of pollution on biodiversity such as downstream water pollution and downwind air pollution can be significant.

Urban stormwater may contain high levels of contaminants such as faecal bacteria, nutrients, chromium, cadmium, lead, nickel, hydrocarbons and chlorinated hydrocarbons. In rural areas, irrigation run-off from farming activities may sometimes contain insecticides, fertilisers and herbicides that have been applied to crops. This run-off may affect aquatic and marine organisms living in the catchment and its associated estuaries and in-shore marine ecosystems.

Sediments and Nutrients

Sediments and nutrients from urban effluent and agricultural chemicals are known to be polluting the inshore reefs of the Great Barrier Reef World Heritage Area, killing coral, encouraging the growth of sessile algae and changing the energy balance and dynamics of marine ecosystems (Wachenfeld et al. 1998). Increased sediment loads lead to muddier systems with less light for bottom communities and disturbance to benthic fauna as a result of siltation. Sediment and nutrient delivery to the Great Barrier Reef from land-based discharges (sources) has increased four-fold in the last 140 years. For the central reef system, 39% of all nitrogen and 52% of all phosphorus originated from river inputs, while sewage discharges accounted for 2.3 and 7.7% of nitrogen and phosphorus, respectively.

Soil loss that results in sedimentation and a reduction in water quality may also affect aquatic and marine biodiversity. Much of the estimated four-fold increase in sediment on to the Great Barrier Reef has occurred during the last 40 years. This loss results from land uses such as grazing, cropping and urban development that are being undertaken on an unsustainable basis (see *Pollution sources on the Great Barrier Reef catchment*). During cyclones, sediment plumes have been recorded at least 100 km offshore. Non-lethal sediment loads may become lethal if the nutrient levels are also elevated (Wachenfeld et al. 1998).

Impacts of Pollutants

Potential effects of pollutants on ecosystems include changes in the abundance of species, interruption to energy and nutrient flows, modification of habitats, reduction in soil, water and air quality, and changes to the stability and resilience of ecosystems. Where species consumed by humans are affected by pollution, there is the potential for serious human health problems. The release of total nitrogen and phosphorus into the environment results in many changes including the likelihood of more frequent and intensive algal blooms in waterways. The dominant release of nitrogen across the Murray-Darling Basin is from unimproved pastures (45% of total aggregate emissions of nitrogen) and cropping (31%), while cropping provides the source of over 50% of phosphorus emissions.

Pollutants can act synergistically to cause uncertain long-term effects on biodiversity. Examples of these interactions can be observed in western Tasmania where acid run-off and acid rain has killed mountain vegetation and affected aquatic ecosystems (e.g. 'dead' sections of river systems such as the Queen and King Rivers). These effects are compounded by the cyclical nature of ecosystem processes, which disperse pollutants widely from their sources and may affect biodiversity at considerable distances in a variety of surprising ways. One reason that mining uranium in Australia remains controversial is the potential effect of these operations on unique environments such as the wetland ecosystems of Kakadu National Park. This is despite two decades of research and monitoring that has not revealed any significant environmental impact on the Park from uranium mining (see *Uranium and biodiversity*).

Uranium and Biodiversity

Uranium is a naturally occurring, radioactive element that is capable of nuclear fission and is used as a source of nuclear energy and for nuclear weapons. Naturally occurring isotopes of uranium have half lives ranging from 4.5 billion years to 250 000 years. About 99% of uranium is in the form of U-238, which has a half life of 4.5 billion years. Mine tailings contain uranium ore with the uranium removed - the remaining radionuclides have a half life of 77 000 years. Uranium and its decay products give off alpha, beta and gamma radiation. This radiation can cause lethal genetic mutations and kill living organisms.

Uranium is mined at three locations in Australia, including the Alligator Rivers Region of the Northern Territory in which lies Kakadu National Park. Surrounded by Kakadu, Ranger Uranium Mine has been operating for 20 years. Uranium mining is controversial from an environmental and biodiversity protection perspective because of the perception of potential accidents that could occur at all stages of the nuclear fuel cycle, including the storage and disposal of various levels of radioactive wastes. Most uranium mining occurs in remote areas of the country that are considered to be of value as wilderness. During the milling of uranium ore, the uranium is removed from the crushed rock and concentrated for transport, but other radioactive substances are left in the residue, which is referred to as tailings. Initially, about 85% of the original radioactivity present in the ore is discarded in the tailings. After several months, the level of radioactivity in the tailings decays to about 70% of that originally present in the ore. The tailings then decay with a half-life of around 77 000 years. Safe management of all stockpiles and mining wastes is essential to prevent release of contaminants and ensure proper protection of the environment.

Uranium mining has attracted considerable attention since 1996 as a result of proposals for new mines (e.g. 11 identified deposits in Western Australia) and the use of remote, arid regions of the country for storage of radioactive waste. For example, during January 2000, the Commonwealth Department of Industry, Science and Resources was conducting drilling and other investigations of six potential sites for storing radioactive waste around Woomera (SA). One concern is that if radioactive waste were to be held at these sites, it may leak into the underground water and find its way into aquatic ecosystems, mound springs and the terrestrial food chain.

The Jabiluka Uranium Mine, 25 km to the north of the Ranger mine in the region of the Kakadu World Heritage Area, is controversial. In October 1996, Energy Resources Australia (ERA) submitted a draft environmental impact statement (EIS) for mining uranium at the site. Following the completion of assessment of the EIS and a Public Environment Report for alternative development options at Jabiluka, the Minister for the Environment found that there did not appear to be any environmental issue that would prevent either alternative from proceeding.

The Minister made environmental recommendations that were subsequently endorsed by the Minister for Resources and Energy. A principal requirement was that all mill tailings would be returned underground to the mine void and to specifically constructed stopes or silos instead of tailing pits as proposed by ERA.

During October and November 1998, a mission from the World Heritage Committee visited the region and site to assess any ascertained or potential threats to the World Heritage values of Kakadu National Park that might arise from the proposal. Wasson et al. (1998), among others, made a submission to the mission specifying environmental concerns arising from weaknesses in the mine design at the mine site (e.g. geomorphology, hydrology and biology). These concerns included the effect that severe weather events may have on the mine storage facilities, leaching of chemicals from the tailings into the ground water, catchment and surrounding wetlands, effects on the aquatic ecosystems of the region, and the effects of climate change given that the tailing storage must be viable for millennia. The mission reported to the World Heritage Committee noting severe ascertained and potential dangers to the cultural and natural values of Kakadu National Park posed by the proposal for uranium mining and milling at Jabiluka and recommended that the proposal should not proceed. Because the Australian authorities had not had sufficient time to respond to the report, the World Heritage Committee made no firm decision on the future status of Kakadu at the November 1998 meeting. The Committee requested that the Supervising Scientist conduct a full review of the areas of scientific uncertainty. The issues specified were hydrological modelling, prediction and effect of severe weather events, storage of uranium ore on the surface and the long-term storage of mine tailings. The Supervising Scientist's assessment of the Jabiluka project was submitted to the World Heritage Committee in April 1999. The overall conclusion drawn was that the natural World Heritage values of Kakadu National Park are not threatened by the development of the Jabiluka mine and that the degree of scientific certainty that applies to this assessment is very high (Johnston & Prendergast 1999).

An independent scientific panel (ISP) was convened by the International Council of Science Unions (ICSU) at the request of the World Heritage Committee to review the report by Johnston and Prendergast (1999). The ISP report was provided to the Supervising Scientist in May 1999. The Supervising Scientist provided a supplementary report to the World Heritage Centre addressing the issues raised in the ISP review. The World Heritage Committee met in July 1999 and resolved not to inscribe Kakadu as 'World Heritage in danger'. After further investigations, the World Heritage Committee met in November 2000 and decided that the mine and mill proposal at Jabiluka does not threaten the natural values of Kakadu National Park.

The National Pollutant Inventory [BD Indicator 20]

The Commonwealth government supports a National Pollutant Inventory (NPI), an Internet database designed to provide the community, industry and government with information on the types and amounts of selected chemicals being emitted to the environment. Australian industrial facilities such as petroleum refineries, chemical manufacturing plants and sewerage treatment plants using more than a specified amount of the chemicals listed on the NPI reporting list are required to estimate and report emissions of these substances annually. The location of these facilities is maintained by the NPI and is largely tied to major human settlements.

Emission sources for an airshed typically include motor vehicles; solid fuel burning; agricultural-related burning, fuel reduction and bushfire controls; and domestic/commercial solvents and aerosols. Industrial discharges into Cockburn Sound in Western Australia have been associated with massive loss of seagrasses and substantial levels of contamination of sediments and fishes.

Research into the Effects of Pollution on Biodiversity [BD Indicator 20]

Persistent pollutants in frogs

The disappearance (perhaps extinction) of frog species in eastern Australia and, indeed globally, is of serious concern. One explanation is the direct and indirect affects of airborne pollutants.

Mann and Bidwell (1999) reviewed environmental toxicology in Australian frogs and noted that few such studies have been carried out on native fauna. They reported that the single largest group of potential chemical pollutants that Australian frogs might encounter are the various pesticides used in agriculture and pest management. Much of the recent work examining the effects of pesticides on amphibians has concentrated on the newer generations of pesticides such as pyrethroids, carbamates and organophosphates, although there has been a resurgence of interest in the older organochlorine insecticides such as DDT because of their persistence in ecosystems and food chains. In particular areas, biological agents such as the chytrid fungi, iridoviruses or predation could also be involved in the decline of frog diversity.

Persistent Pollutants in Crabs

In coastal Queensland, Mortimer (1999) quantified the trace metals, metalloids and pesticide content in intertidal Burrowing Crabs (*Australoplax tridentata*) and the large Mud Crab (*Scylla serrata*). Estuaries between Cairns and Brisbane were sampled and residues of dieldrin were found at all locations, and heptachlor epoxide and DDT were recorded at most. Calculations of ambient exposures to organochlorines based on residues in crab tissues indicated that dieldrin exceeded national water quality guidelines for protection of aquatic ecosystems at all sampling locations, but exposure to DDT and its metabolites was below the threshold of concern. Use of DDT, dieldrin and heptachlor is banned in Australia.

The primary former uses of dieldrin included treating crops for the control of root fly larvae, locusts, crickets and grasshoppers; in building and industry to control termites; and to control disease vectors such as cockroaches, fleas and mosquito larvae. Use of dieldrin was progressively restricted from 1973 and banned in June 1994. Heptachlor was used as a soil treatment to control ants and grubs in sugar cane areas; the Banana Weevil Borer (*Cosmopolites sordidus*) in banana plantations; and to control termites in buildings and other structures. Agricultural use of heptachlor ceased in 1987, but it was still used for termite control in Queensland until June 1995. DDT was used extensively in agriculture to control both crop and livestock pests. It was also used for domestic control of fleas, lice, mites and lawn grubs. Domestic uses were banned in 1973, and agricultural use was progressively restricted until it was banned in 1987. Dieldrin, heptachlor and DDT are extremely hazardous to humans and biodiversity.

Oil Spills

More than 22 million tonnes of oil is shipped in Australian waters each year as tanker cargo, including many ships that pass through or near sensitive marine environments such as the Great Barrier Reef. Although Australia has experienced relatively few oil spills compared with elsewhere in the world, several major incidents during the 1990s (e.g. 1991: *Kirki*, 17 900 t of oil; 1995: *Iron Barron*, 325 t; 1999: *Laura D'Amato*, 80 t) are a reminder of the need for suitable monitoring of shipping standards and transits and the need for a high quality emergency capability. The probability of one or more major spills from tankers could be as high as 37% in any five-year period, and 84% in 20 years (Bureau of Transport and Communications Economics 1999).

Ocean Dumping

Since 1975, an international agreement known as the London Convention (formerly called the London Dumping Convention) has controlled sea dumping internationally. In order to ratify the London Convention, Australia enacted the *Environment Protection (Impact of Sea Dumping) Act 1981* which prevents the dumping of some wastes and provides for the regulated dumping of other substances in waters off Australia and its External Territories such, as the AAT, Heard Island and McDonald Islands, Macquarie Island, Norfolk Island, Cocos (Keeling) Islands and Christmas Island. Most permits issued are for the creation of artificial reefs or for disposal of uncontaminated dredge spoil.

5

Exotic Species and Genetically Modified Organisms

INTRODUCTION

One of the objectives of the National Strategy for the Conservation of Australia's Biodiversity was to implement effective controls for at least one introduced mammal and at least three introduced plant species by the year 2000. The National Weeds Program and the National Feral Animal Control Program aim to provide a strategic framework for addressing such goals. The Commonwealth EPBC Act also includes provision for protecting Australia's environment from invasive species. However, as indicated below, the magnitude of the task is daunting.

Pest animal species

Australia's pest animals are species that cause severe damage to natural and agricultural systems (Olsen 1998). These pest animals are either domestic animals that have gone wild or those that were introduced for the control of pests or for recreation. Pest animals causing most public concern include the European Rabbit (*Oryctolagus cuniculus*), European Fox (*Vulpes vulpes*), cat (*Felis catus*), pig (*Sus scrofa*), goat (*Capra hirus*), donkey (*Equus asinus*), camel (*Camelus dromedarius*), Water Buffalo (*Bubalus arnee*), Mosquito Fish (*Gambusia holbrooki*), Northern Pacific Seastar (*Asterias amurensis*), European Carp (*Cyprinus carpio*) and Cane Toad (*Bufo marinus*). Pest animals may damage vegetation and soils, foul water or compete with native animals for habitat and food.

About 20 species of mammals, 25 species of birds, several amphibians and 19 species of freshwater fish make up the pest vertebrate populations of Australia. TAPs have been prepared under the National Feral Animal Control Program for four of these species: the European Fox, cat, rabbit and goat. These plans focus on strategic approaches to reducing, to an acceptable level, the effects of processes that threaten the long-term survival of native species and ecological communities.

Introduced species can have a major effect on offshore islands such as Lord Howe Island and Macquarie Island. The endangered Woodhen (*Tricholimnas sylvestris*), one of the world's rarest bird species, is found on Lord Howe Island. During the 1900s, the population of Woodhens has declined significantly in numbers as a result of disturbance by feral animals, hunting by humans and habitat loss. Thirty-four plant and animal species have been introduced on Macquarie Island, some deliberately for human consumption. The list ranges from domestic animals (which generally did not survive the harsh conditions) to insects such as the flea, which was probably introduced with the Black Rat. The greatest environmental damage on the Island has come from introduced mammals like cats and rats that eat ground-nesting seabirds, eggs and chicks or from rabbits, which eat the vegetation and contribute to soil erosion. More remote islands such as Heard and McDonald Islands are considered free from introduced species, although introductions of microorganisms could have occurred.

Clarke et al. (2000) provided detailed descriptions of over 50 environmentally significant vertebrate and invertebrate pests in freshwater, terrestrial and marine environments. Eight of the non-native pest animals were from freshwater habitats, over 30 from terrestrial habitats and at least 10 from marine habitats. Distribution maps have been produced for each species as well as total number of pest species per IBRA, drainage basin and IMCRA region. These show that the highest numbers of pest species on both an IBRA, drainage basin and IMCRA basis occur in south-east Australia where the human population is highest.

INVERTEBRATE SPECIES AND MICROORGANISMS

About 500 introduced invertebrate species are thought to be changing the environment, including the European Honeybee (*Apis mellifera*) that may out-compete smaller, typically solitary Australian native bees (Paton 1996). The management of species such as the European Honeybee, which have been introduced for commercial purposes but have subsequent effects on native species, brings considerable challenges. This has also been observed with grasses introduced for pastoral purposes. The potential for the introduction of microorganisms has not been systematically explored, but could have a significant effect on biodiversity. This could be of particular importance in remote environments such as Antarctica, where it is thought that the introduction of microorganisms is likely to have occurred.

New species continue to appear such as the recently discovered Black Striped Mussel (*Congeria sallei*), an invasive freshwater mussel from Central America, which was found in Darwin Harbour. Given the tolerance of this mussel to a range of temperature and salinity, it could potentially infest all Australian ports. A major program was put in place to eradicate the mussel, focusing on the application of large quantities of chlorine and copper sulfate. While the mussels are thought to be unable to survive this treatment, there is a strong chance that the species could become re-established and a national task force has been formed to combat the pest.

As so little is known about which marine species normally inhabit most ports and harbours, it is almost impossible to record the rate of introductions.

When Least Expected

Some introduced animals that have been around for decades can also become pests. For example, the Crazy Ant (*Anoplolepis gracilipes*) was introduced to Christmas Island in the 1930s, and only recently has become a major threat to biodiversity. Species introduced to control pest species can also become pests themselves. The introduction of the South American Cane Toad (*Bufo marinus*) into Queensland in 1935 from Hawaii to control two insect pests of sugar cane was a failure. The insects were later controlled using insecticides and other management practices. Meanwhile, the Cane Toad continues to thrive relying on a range of prey including native invertebrates and small vertebrate species, and they are also thought to compete with native frogs and snakes for these foods. In addition, the toad has toxic skin secretions that are known to kill vertebrate predators such as goannas, quolls, skinks and snakes. The toads have been spreading, mainly to the north, since their introduction. In December 1998, they were noticed around 500 km south of Darwin. They have been found recently in as far north as the Koolatong River in east Arnhem Land to just inside the south-east corner of Kakadu National Park and throughout the Katherine River System.

THE THREAT TO BIODIVERSITY ON CHRISTMAS ISLAND FROM THE CRAZY ANT

The Crazy Ant (*Anoplolepis gracilipes*) has become a major threat to biodiversity on Christmas Island. It is a tropical species native to western Africa and was first found in Australia on Christmas Island in the 1930s. It has become a pest only recently there, but is already seen as a direct threat to as many as 20 animal species, including sea birds, land birds, mammals, reptiles and three species of crabs. Millions of the Red Crabs (*Gecariodea natalis*), for which Christmas Island is famous, have been killed by the Crazy Ant. The crab is critical to the dynamics of the unique rainforest communities on the island, so its elimination has a direct effect on forest structure and processes. The presence of Crazy Ants can also lead to canopy dieback due to the mutual relationship with scale insects that produce honeydew (scale insects cause a disease that can result in dieback; honeydew is a food source of the ant and the ants protect the scale insects from other predators). Several reasons underly the major threat imposed by these ants: their extremely broad diet, the formation of supercolonies with extraordinarily high and sustained densities, and they forage in three dimensions. Contact sprays, dusts and toxic baits have been successfully used to control the ant, and because there appears to be no native ant species on Christmas Island, biological control could be feasible long-term. Unless host specificity could be guaranteed, this method would not be suitable for mainland Australia, where the Crazy Ant has recently been recorded in the north-east of the Northern Territory.

Environmental weeds [BD Indicator 3.2]

Environmental weeds are plants that invade natural ecosystems and can cause major modification to indigenous species and ecosystem function (as opposed to horticultural or agricultural weeds, although a species may be both). They are considered one of the greatest threats to nature conservation in Australia (Commonwealth of Australia 1999; Low 1999), having been implicated in the extinction of four plant species (Groves & Willis 1999). The most recent estimate of the number of environmental weeds in Australia is 1060 (Swarbrick & Skarrat 1994). Virtually all aquatic and terrestrial vegetation communities in Australia are affected by environmental weeds. There are, however, considerable differences in the level of invasion, both within and between vegetation types, with disturbance history and proximity to human development being key factors.

One species of great concern, the Rubber Vine (*Cryptostegia grandiflora*), entangles trees and other vegetation and eventually smothers them. The vine is spreading at an alarming rate through the river systems of southern Cape York, the Queensland part of the Gulf of Carpentaria and along the coast as far south as the Burnett River near Bundaberg, destroying the riverside vegetation in these regions.

For aquatic weeds, patterns of rainfall in particular have a major influence on their distribution and management. A network of people responsible for aquatic weed management is needed to gather information in order to assess the aquatic weed situation annually across Australia. Without this, it is impossible to describe the extent of effects at a national level, although several case studies for individual species are available.

Indigenous Species

Not all environmental weeds are introduced from other countries. They can be indigenous species that have been deliberately planted for horticulture beyond their natural range or where altered disturbance regimes encourage their spread into areas where they did not grow previously. Despite being 'native' plants, weeds of indigenous origin can severely disrupt ecosystems. Examples are Coastal Tea-tree (*Leptospermum laevigatum*), Cootamundra Wattle (*Acacia baileyana*) and Sweet Pittosporum (*Pittosporum undulatum*) (Mullett 1996).

Effects of Environmental Weeds

Environmental weeds can have a range of effects on natural systems, although quantitative studies of these effects are uncommon (Adair & Groves 1998). Despite the paucity of studies, however, it is likely that major changes to ecosystem structure will cause consequent losses of biodiversity. This has been observed in New Zealand, where several species introduced deliberately for naturalisation in the wild have dispersed effectively from the locations in which they were planted, reducing the diversity of both plant and animal species in all cases (Williams & West 2000). The effects of environmental weeds are not always detrimental. For example, they can provide some native animals with additional food sources and shelter. In these cases, management strategies need to reflect the supply of alternative shelter and food sources. Equally important, management options need to include consideration of the control of native dispersal agents such as birds and mammals. Weeds can be spread widely by animals, as demonstrated for species such as the Brushtail Possum.

Pasture Plants

Plants legally introduced into Australia for pasture production or horticulture have the potential to become environmental weeds.

Lonsdale (1994) demonstrated that greater than 99% of species introduced between 1947 and 1985 that were considered useful for pasture production caused weed problems in both cropping and conservation areas in northern Australia. Only 5% became useful for pastoral use. Mitchell and Williams (2000) described two vigorous wetland grasses introduced for ponded pastures in Queensland in the 1980s which outcompeted native plants and have drastically altered the habitat for many waterfowl. Consequently, the net benefit for Australia of introducing such species needs to be seriously considered before similar species are introduced.

An introduced pasture species that is quickly becoming a major environmental weed in central Australia is Buffel Grass (*Cenchrus ciliaris*). This is a perennial grass, which has been shown to reduce the diversity and abundance of both native vascular plants and invertebrates (Pitts & Albrecht 2000). Seed of this species is spread by helicopter by some pastoralists and once established it is a difficult species to control. Burning is not an effective long-term solution as the grass recovers rapidly and completely, and suppresses the germination of native plant species.

Sleeper Weeds

Many of the introduced plants already in Australia may become weeds given sufficient time or a change in conditions (e.g. disturbance, introduction of a pollinator and climate change). 'Sleeper weeds' (those species that have naturalised, but not yet exponentially expanded their populations) are a major concern.

Groves (1999) identified two ecological factors that were useful in predicting sleeper weeds: time from naturalisation, and relocation to a more favourable site. Some progress has been made, with recent studies identifying 'potential environmental weeds' (non-native species only) that are amenable to eradication (Csurches & Edwards 1998). There is still, however, no effective response mechanisms in place to retrospectively apply weed risk assessment to naturalised species that can still be controlled.

Nursery Plants

Another source of concern is garden plants that can become potential environmental weeds. In 1999, the Nursery Industry Association identified 860 invasive plants available through nurseries. To minimise the impact of this practice, the current emphasis is on education and raising awareness, with the government, nursery industry and horticultural media working together to find solutions (Blood 1999). Currently, 50 of the plants have been voluntarily withdrawn from sale. Control of this trade is made more difficult by the complex and inconsistent approach to noxious weed legislation in Australia (Thorp & Lynch 1999).

Policies and programs for weed management

At the national level, the developing awareness during the 1990s of the threat of weed species to both production and conservation systems achieved formal recognition with the launch of the National Weeds Strategy in mid-1997.

A central component of the strategy was the identification of the most serious weed problems in Australia, which resulted in an official list of 20 Weeds of National Significance (WONS) (Thorp & Lynch 2000). This was developed on behalf of Agricultural and Resource Management Council of Australia and New Zealand (ARMCANZ), Australia and New Zealand Environment Conservation Council (ANZECC) and the Ministerial Council on Forestry, Fisheries and Aquaculture (MCFFA) and has been agreed with the states and territories of Australia after extensive consultation.

The final 20 species were selected from an original list of 71 species based on their invasiveness, economic, environmental and social effects, current distribution, potential for spread and effect in reducing the growth of desirable plants. Whereas all 20 species were identified as a threat to plant communities and endangered plant and animal species, six of the weed species were classified as primarily a threat to the environment, another five as primarily a threat to primary production systems, and nine weeds have both environmental and primary production effects. Maps of the current and predicted distribution of the 20 species have been produced and a major publication on the WONS has been released (Thorp & Lynch 2000). While endorsement of the list is not accompanied by any specific form of guaranteed funding, it will assist the Commonwealth, states and territories in determining funding priorities as well as providing direction to research organisations, commercial partners, industry and community groups.

National coordination teams are being established for each of the weed species and individual management plans prepared. The implementation of the plans has been slower than expected due to the need to engage a range of stakeholders in their development. The intention of the plans is to focus on strategic management and on species that have national priority.

Considering the number of environmental weeds in Australia, and the potential threat posed by 'sleeper weeds', there is some concern that focusing resources on a small number of nationally significant species may not be the best approach. To help complement this strategy, an 'alert list' has been developed of 28 non-native plant species that are in the early stages of establishment and have the potential to become a significant threat to biodiversity if they are not managed. Funding priority for their management or control will be given to projects that promote the long-term protection of remnant vegetation, especially ecologically threatened communities, and that target isolated populations of the weeds to prevent further establishment and expansion of the weed.

Diseases

Introduced diseases pose a major threat to biodiversity in Australia, as shown by the effect of dieback disease, caused by the fungus *Phytophthora cinnamomi.* This species was introduced into Australia in the 1800s, probably accompanying plant material introduced for agriculture. Dieback disease can result in great changes in a broad range of ecosystems including heathlands, shrublands, woodlands and forests. About 20% of the 9000 vascular plants of south-west Western Australia may be at risk from dieback disease, including more than 80% of all Proteaceae which are particularly susceptible. At least 10% of the remaining jarrah forest in south-west Western Australia is infected with dieback disease, although this is considered an underestimate. Changes in vegetation communities associated with *Phytophthora* would affect fauna, especially for species such as the Honey Possum (*Tarsipes rostratus*), which rely on a specialised diet of pollen and nectar mainly from highly susceptible proteaceous plants.

Phytophthora is also a problem in eastern Australia, and has been found in the highlands of Tasmania. The flora of northern Australia is also considered susceptible to *Phytophthora*, so several hygiene measures have been put in place to stop its introduction via potted plants brought in for horticultural purposes or use in gardens.

The extent of the threat of *Phytophthora* to Australia's species and ecosystems is recognised in the 1996 *National Strategy for the Conservation of Australia's Biodiversity* where it is the only pathogenic taxon specifically cited. The Commonwealth EPBC Act also lists the disease caused by *Phytophthora* as a key threatening process. A nationally coordinated threat abatement process is being developed to manage the effect of *Phytophthora* dieback on Australian ecosystems. A draft plan was circulated in July 1999 which estimated that over $10 million would be needed to be spent over five years to implement the plan.

Disease has also been linked to the decline of some frog populations in Australia. There is strong evidence that a chytrid fungus is associated with the death of some frogs, but it is not known if it is the direct cause of mortality or an indication of other environmental stress. Other diseases already in Australia may become a threat, in much the same manner as 'sleeper weeds' and pests.

Quarantine

Effective quarantine is a key factor in minimising the introduction of potential weeds, pests and diseases from both other countries and within Australia. The recently introduced weed risk assessment was explicitly designed to detect weeds of both environmental and agricultural concern (Steinke & Walton 1999). This system has been endorsed by a wide range of client groups and is now used on all new plant imports whether they enter Australia as seeds, nursery stock or tissue culture, regardless of their use in Australia.

Managing Introduced Species

Integrated management

Integrated management involves coordinated use of various control techniques, and integrating control with other activities. The major aims of integrated management for environmental weeds are to (Vranjic et al. 2000):

- effectively contain the spread of existing weeds
- manage the environment to prevent the incursion of new weeds
- rehabilitate the disturbed ecosystem as far as possible.

In natural systems, this involves a range of considerations such as the use of herbicides (and their potential effects on non-target species), the appropriate situations where fire can be used and the importance of passive and active revegetation.

Biological Control

Biological control is one of the tools used to manage a range of introduced organisms. The *Biological Control Act 1984* is the principal enabling legislation at the Commonwealth level. Biological control is not always the 'magic bullet', particularly for terrestrial weeds with large soil seed reserves (Briese 2000). Consequently, biological control is more frequently seen as part of an overall management strategy.

One of the critical tests for new organisms being considered as biological control agents is whether they will affect other non-target organisms. Stringent tests are now enforced to help ensure that biological control agents are effective and do not become pests themselves. However, the system is not fail-safe as illustrated by the untimely escape in 1995 of rabbit calicivirus (RCD) from Wardang Island in South Australia.

RABBIT CALICIVIRUS AND OTHER MANAGEMENT TOOLS

Agents such as RCD, and its precursor myxomatosis, are essential management tools, given the number and extent of rabbits in Australia. RCD is highly infectious to rabbits and can cause mortalities of up to 95% among adults, although regional variation in infection rates can be significant.

While an initial reduction in rabbit populations may be brought about by an effective outbreak of viruses such as RCD or by poisoning or drought, to keep populations low, follow-up techniques such as poisoning, warren ripping and fumigation are recommended. At another level, rabbit control may need to be integrated with the control of other pest species such as foxes and feral cats. There has been some concern that the reduction in rabbit numbers may lead to predators switching their prey to native animal species.

Western Shield

The Western Shield Program, managed by CALM in Western Australia, aims to bring at least 13 fauna species back from the brink of extinction by controlling foxes and feral cats, on almost five million hectares of land.

The main weapon in the fight against the fox and feral cat is use of the naturally occurring poison 1080, found in native plants called gastrolobiums ('poison peas'). Although native animals in south-west Australia have a high tolerance to the poison, introduced animals do not. In the south-west forests, scientific research and monitoring has shown that where fox numbers have been reduced by baiting with 1080, there has been a dramatic increase in native animal numbers.

Endangered native animals such as the Numbat (*Myrmecobius fasciatus*), Noisy Scrub-bird (*Atrichornis clamosus*) and Ring-Tailed

Possum (*Petropseudes dahli*) are increasing in numbers after being reintroduced into forest inside and outside reserves. Already, three forest-dwelling mammals, the Woylie (*Bettongia pencillata*), Quenda or Southern Brown Bandicoot (*Isoodon obesulus*) and Tammar Wallaby (*Macropus eugenii*) have been removed from the State's Threatened Fauna List and the status of other species has been improved as a result of successful conservation management under the Western Shield Program.

Community Involvement

The successful management of pests and weeds requires the informed and motivated participation of people throughout Australia. An important element of the success of the Western Shield Program described above is the cooperation and support of local communities. Many private landowners and Land Conservation District Committees have helped with fox baiting by laying baits on their own land where it is next to conservation reserves and state forests. As well as government funding, the baiting program is sponsored by Alcoa Australia, Cable Sands and Westralian Sands.

The Ballarat Region Gorse Task Force in Victoria is another example that involves a range of groups in the management of an environmental weed.

BALLARAT REGION GORSE (*ULEX EUROPAEUS*) TASK FORCE

Gorse (*Ulex europaeus*) is a serious environmental and agricultural weed in south-east Australia because of its invasiveness and the difficulty and expense involved in conventional control. It invades bushland, reducing access and conservation values and threatening the survival of rare and endangered plants. On pastoral land, gorse significantly reduces pasture and animal productivity. Gorse also provides habitat and shelter for vertebrate pests such as rabbits and foxes, and increases fire hazards. In Victoria, gorse is common in the central highlands, the south-west and parts of Gippsland.

The Gorse Task Force Area is centred on Ballarat and covers some 800 000 ha of urban, farming and forested land in the central highlands of Victoria.

The Ballarat Region Gorse Task Force represents 43 Landcare groups across the central highlands of Victoria. The Task Force has prepared a Gorse Control Strategy which targets infestations along roadsides and waterways where the greatest potential for spread exists, and where there is a demonstrated public benefit associated with control.

The Task Force approach is based on a model with strong partnerships between landholders and Landcare groups, local government, catchment management authorities and the Department of Natural Resources and Environment (DNRE). The model empowers community groups to lead and coordinate the implementation of agreed community-based weed control strategies. The role of DNRE is to support these strategies through providing community education, extension and enforcement programs to ensure that a lack of action by a minority does not jeopardise the good work of most of the community.

The Gorse Task Force has secured funding for implementation of the Gorse Control Strategy, including the employment of facilitators and provision of incentives for onground control where there is a demonstrated public benefit. DNRE have provided compliance officers, funded through their 'Enhanced Enforcement' program.

The Strategy identifies facilitation as the key to raising community awareness of the problems caused by gorse, for coordinating strategic programs for long-term management, and most importantly, for creating community acceptance of responsibility for gorse control.

Funding and research into exotic organisms

Environmental weeds: An indication of the nature of the research on environmental weeds can be gained through the annotated bibliography developed by Swarbrick and Timmins (1997). At the national level, the main focus of research has been through two CRCs: Tropical Pest Management in northern Australia and Weed Management Systems in southern Australia, established in 1995. The former CRC ceased operation in 1998, but individual organisations are still working on the ecology of environmental weeds in the north.

The Weed Management Systems CRC has worked on seven 'key' environmental weeds: Bitou Bush (*Chrysanthemoides monilifera* ssp. *rotundata*), Boneseed (*Chrysanthemoides monilifera* ssp. *monilifera*), Bridal Creeper (*Asparagus asparagoides*), Broom (*Cytisus scoparius*), Blackberry (*Rubus fruticosus*), Horehound (*Marrubium vulgare*) and St John's Wort (*Hypericum perforatum*).

In 2001, the CRC for Weed Management Systems was granted a further seven years of funding. This new CRC for Australian Weed Management has two programs of relevance to environmental weeds. The 'Landscape Management' Program will focus on the management of key habitats to control weeds, the management of model weed types and enhanced biocontrol strategies. A 'Weed Incursion and Risk Management' Program is also being developed that will examine risk assessment and management of sleeper weeds.

At the national level, the main source of funding for environmental weeds is through the National Weeds Program. From 1996 to 2002, a sum of $28.5 million has been allocated to this Program. In May 2001, the Commonwealth government announced funding through this Program for onground projects related to the management of established weeds that could pose a significant threat to biodiversity (as identified in the 'Alert list' of 28 species).

Pest Animals

The CRC for Biological Control of Pest Animals started in 1999 and builds on the work of the CRC for Vertebrate Biocontrol. The research focus of the CRC is to develop ways to control populations of the rabbit, fox and introduced house mouse through using vaccines that cause sterility. Before sterilisation for these animals becomes a reality, many ecological hurdles will need to be identified and overcome.

The total resources involved in the CRC for biological control is estimated to be around $57 million over seven years. The Centre has 20 postgraduate students and the equivalent of 51 full-time research staff. At the national level, $18.9 million was allocated to the National Feral Animals Control Program between 1996 and 2002.

Genetically Modified Organisms

Distribution and abundance of genetically modified organisms

A GMO is any organism with genetic material that has been altered by genetic engineering. GMO research programs have much in common with traditional plant and animal breeding programs whose intention is to produce individuals with new genetic composition, better adapted to the needs of agriculture, medicine or some other productive use.

GMOs are novel because the tools of molecular biology allow genes to be introduced into species that would be difficult or impossible using traditional breeding techniques. There are potential ecological benefits from genetically modified (GM) plants and animals, including the prospect of plant varieties to rehabilitate salt-affected areas or soils contaminated by heavy metals. Some potential benefits are indirect and difficult to predict, such as reductions in the use of pesticides (Barnes 2000).

Like other introduced organisms, there is also the potential for ecological costs. With the proliferation of GM products and the almost exponential growth of land use for GM crops (AAS 1999), there is growing interest in the estimation of the risks to the environment of exposure to such products. The major potential hazards for biodiversity posed by GM species may be summarised under four headings (Levin 1992):

1. Transgenic leakage into related wild populations may occur by hybridisation, where genes from the GM species move by cross-pollination into wild relatives of the same or different but closely related species
2. GM populations may become invasive of natural habitats, competing most intensely with wild relatives
3. GM species may be a direct hazard to non-target species, where the specific properties of GMOs (e.g. insect resistance in crop plants) may make them allergenic or toxic to a range of species that are part of the ecosystem shared by the GM species
4. GM species may have indirect effects through changes in agricultural practice, including establishment of crops and livestock in areas considered marginal for agriculture, made possible by the novel properties of GMOs.

The introduction and spread of GMOs in Australia is regulated by the Interim Office of the Gene Technology Regulator and its Genetic Manipulation Advisory Committee (GMAC). Institutional Biosafety Committees (IBCs) operate in institutions where GMO research is undertaken.

They oversee the development of new proposals and the conduct of research, acting as a first filter before new ideas are forwarded to GMAC for their consideration. General releases of GMOs are controlled by the Commonwealth Minister for Health and Aged Care under interim arrangements announced by the Commonwealth in August 1999. A national system of statutory regulation will be in place by July 2001.

Because of the prospect of the spread of genes from GM crops to adjacent non-GM crops, or between related species by hybridisation, buffers between field trials and adjacent wild and non-transgenic crop populations are used.

The extent and utility of these buffers has been the focus of some research, with quantification of some aspects of genetic systems, dispersal and gene introgression. However, there has been little work on comprehensive, quantitative modelling of ecological risks of GMOs (Timmons et al. 1996; Hails 2000) in Australia or elsewhere. Small-scale proposals include laboratory, glasshouse or clinical applications of recombinant DNA under contained conditions. They do not involve the release or cultivation of individuals under field conditions. A total of 4811 small-scale proposals were assessed by GMAC between 1981 and June 1999. In June 1999, there were 1681 active, small-scale, contained projects.

Deliberate releases of GMOs involve trials under field conditions and range from areas smaller than 1 ha up to many thousands of hectares, and numbers of individuals ranging from fewer than 50 to many millions. In June 1999, there were 13 large-scale projects and 109 deliberate releases underway, the latter an increase from June 1998 of 69 projects. Most deliberate releases are for commercial crops. Cotton (*Gossypium* spp.) and Canola (*Brassica* spp.) have been the subjects of most attention and make up the bulk of field trials, although more than 30 different kinds of GM species had been trialed up to June 1999. GMOs have been trialed in all Australian states. A general release implies that the product is commercially available. There were two general releases in Australia before March 2001, for cotton and carnations.

GMAC assessed almost all proposals as acceptable, after modification and with different levels of protection being required. Two proposals were rejected, both relating to a project focused on the development of bacteria in the guts of ruminant animals that would allow them to tolerate fluoroacetate poisons that naturally occur in native plant species.

One of the concerns was the potential for the bacterium to find its way into the guts of non-target species, perhaps resulting in increased effects on natural vegetation from feral animals. In addition, it is possible that the effect of 1080 bait used in the Western Shield Project in Western Australia could be reduced if feral animals developed tolerance to these poisons.

There are several crop genera where Australian native species are found in the same genus. Of the species listed, GM varieties of potatoes, tomatoes and cotton have been trialed in the field. Many other crop species have the potential to hybridise with native species outside their genus. For example, there are many Australian species in the family Brassicaceae that could hybridise with crops of Canola and Indian Mustard (genus *Brassica*). The native species are important in their own right, and represent important reservoirs for genetic resources for crop improvement. These species may be at risk from unwanted hybridisation, both from genetically modified variants and from stock developed by more traditional means. The risk of hybridisation depends on many factors including pollen size and mode of dispersal, seed dispersal mechanisms, timing of flowering, wind direction, water flows, outcrossing rates and the spatial proximity of populations.

6

Protecting Biodiversity

GENETIC DIVERSITY

Conserving genes and genetic diversity in populations may be achieved through the maintenance of genetic structure within and among subpopulations, or among sets of populations with common evolutionary histories. This strategy improves the ability of populations to adapt to novel environmental conditions and avoids the potentially negative effects of inbreeding (Frankham et al. 2001). Genetic information may also be used to guide the management of captive populations, to set conservation priorities by identifying unique populations, to provide information on population dynamics and mating systems that would otherwise be unobtainable, and to understand the consequences of historical events, such as range expansions, fragmentation and bottlenecks (Moritz 1995).

Conserving Quantitative Genetic Variation

Monitoring the potential for adaptation to environmental change might be well served by an explicit focus on monitoring changes in quantitative environmental variation, particularly the component that is available for evolution, namely, heritable genetic variation.

However, studies that measure overall levels of genetic variation within and among populations do not provide any information on genetic variation of fitness traits that are under selection in the field. The focus on maintaining variability among such populations raises the real threat that the adaptive potential of larger but less distinct populations is being minimised. A rapidly changing environment where the maximisation of genetic variation for traits associated with stress and disease resistance will be critical for long-term persistence.

Alternate approaches for maintaining appropriate genetic variation in populations have been advocated. These include shifting the focus to phenotypic variation of traits associated with stress and disease resistance that will have the effect of selection of favourable genes at many loci (Woods et al. 1999). Another approach is to concentrate on populations that persist at ecological margins as these will have been selected for environmental stress resistance and are likely to have higher frequencies of stress resistance genes than population inhabiting more favourable environments (Hoffman & Parsons 1997). Marginal populations may have evolved generalised stress resistance mechanisms that will allow them greater resistance in stress environments other than those in which they have been selected.

Management strategies should aim to conserve species across a broad range of climatic regions and to conserve all races, variants and subspecies. This will ensure that any genotypes fixed because of local adaptation will be conserved and available to counter future climatic changes.

Population size, numbers and physical isolation

There are insufficient data for this indicator to be a reliable reflection of any general or specific trends in the distribution of genetic variation as a response to changes in the environment.

CHALLENGES ASSOCIATED WITH GENETIC INDICATORS

The entire genome of an individual cannot be measured. The closest science has come to achieving this goal is the Human Genome Project. This involved enormous human and financial resources, sequenced various parts of different humans, not of one individual alone, and was a one-off project. Even if it were possible to measure an entire genome, we would not know which parts of the genome had adaptive significance. The kinds of genetic data that may be acquired reflect the different regions of the genome from which they are sampled. Different regions have different characteristics and the data taken from them may have different applications.

Overall, the recommendations on genetic indicators by Brown et al. (1997) and Saunders et al. (1998) may generate some useful statistics for monitoring species but most current genetic studies of Australian species do not provide sufficient information for the relevant variables to be calculated. Allelic richness and gene diversity measure the basic raw material available for evolution: allelic richness is more sensitive a measure, but is more susceptible to sampling artefacts. The observed heterozygosity per individual (H_o) is a good indicator of processes such as mating systems, and may predict fitness. It is also possible to estimate levels of inbreeding from patterns of heterozygosity. The concepts of allelic richness, gene diversity and observed heterozygosity are explained in greater detail.

Interspecific and intraspecific comparisons of allelic richness are compromised, however, because of biases in the selection of markers. For example, comparisons between taxa and between times will be affected by the choice of markers, because different marker systems produce different estimates of genetic diversity (Moran et al. 2000). Brown et al. (1997) recommended identifying a set of species for detailed monitoring, providing baselines for assessing future changes in genetic diversity. Their recommendations depend on the identification of indicator taxa, and resampling of species genomes to ensure sufficient sample sizes to reliably detect trends in the distribution and amount of genetic variation.

MEASURING GENETIC VARIATION FOR STATE OF THE ENVIRONMENT REPORTING

Allelic richness: Allelic richness is defined as the number of alleles in a sample, standardised for sample size. Two statistics are relevant to allelic richness, the percentage of polymorphic loci (*P*), and the average number of alleles per locus (*A*). To compare allelic richness between species, or to compare samples from the same species at different times, it would be necessary to sample the same markers on each occasion, or to take a random sample of all possible markers. But methodological conventions are to survey available markers and to select informative ones for analysis and reporting. Selection is often haphazard but is rarely random (Glaubitz et al. 1999). In addition, loci are treated as polymorphic only if the frequency of the most frequent allele is less than 0.95, or sometimes less than 0.99 or 0.90. Typically, the number and identity of monomorphic markers are not published, sometimes by editorial policy, if not by convention.

It is also possible to estimate the number of alleles per polymorphic locus, which may be readily compared across studies. In addition, planned and coordinated monitoring strategies could overcome biases through appropriate marker choice.

Gene diversity: Gene diversity, H_e, is the expected level of heterozygosity in a randomly mating population, given observed allelic frequencies. In species with subdivisions (populations), the diversity may be partitioned into within-population (H_s) and between-population (G_{ST}) components. This information may be used to examine patterns of population structure and differentiation and non-random mating.

The level and type of information obtained from genetic studies depends on the markers chosen. Gene diversity is directly available in studies that use codominant markers (including restriction fragment length polymorphisms, RFLPs, isozymes and microsatellites). Studies using dominant markers (e.g. polymerase chain reaction-based markers including amplified fragment length polymorphisms and random amplified polymorphic DNA, RAPD) or haploid markers (e.g. mitochondrial DNA) usually report genetic distances. These studies do not provide any direct measure of gene frequencies. Mitochondrial DNA gives a measure of diversity equivalent to H_e (haplotype diversity) but no measure of inbreeding. This occurs because the various methods used to score genetic markers detect different proportions of the total variation in DNA base sequences. Moreover, the rates of mutation vary widely between different classes of marker. For example, isozymes and RFLPs have low mutation rates and microsatellites have very high mutation rates (Ennos 1996).

Observed heterozygosity: Observed heterozygosity, H_o, is the proportion of heterozygotes in a population, averaged over all loci, and is theoretically available in all studies that use codominant markers. However, this statistic is rarely reported in the literature. Most studies are concerned primarily with differentiation between populations in a single study, so that comparisons between times and between species are rarely considered. Microsatellite studies contrast with studies that use other codominant markers. Most microsatellite studies report both H_o and H_e, allowing computation of inbreeding coefficients (e.g. Baker et al. 2000; Lebas & Spencer 2000; Miwa et al. 2000).

MONITORING ROUND-LEAF HONEYSUCKLE (*LAMBERTIA ORBIFOLIA*)

Coates and Hamley (1999) measured the genetic variation within and between populations of *Lambertia orbifolia*, a large woody shrub restricted to seven populations in Western Australia. They found 12 out of 19 isozyme loci were polymorphic. Levels of genetic variation were roughly comparable with other long-lived woody shrub species endemic to the region. Genetic divergence between population groups was very high, with 44% of variation in genetic diversity being composed of between population differences. They estimated rates of gene flow between the populations to be very low and found consistently low levels of outcrossing. Coates and Hamley (1999) recommended that one population be recognised as a separate conservation unit, on the basis of the degree of genetic differentiation. They reported *A*, *P*, H_e and H_o, together with standard errors for each statistic, for each of the populations, making it possible to resample these populations in future years, to measure trends in diversity.

Species Diversity

Species richness

The most recent estimate of global species richness (Lawton & May 1995) puts the total number between five and seven million species globally, with an upper limit of perhaps 15 million species. Estimates of the total number of species in Australia vary considerably. SoE (1996) reported around one million species, whereas Horwitz et al. (1999) believe that it is likely to exceed ten million. These figures demonstrate the large gaps in knowledge about even the number of species found in Australia.

The ABRS provided the most recent summary of our knowledge of species diversity in Australia. Only a few vertebrate groups are thought to be fully described: frogs, birds, lizards, snakes, birds, mammals and lancelets. Consequently there are very few examples where the number of species is believed to be the final total. Of the groups that are described, certain taxa in Australia contain a globally significant number of species (e.g. reptiles, ants, lichens), as do certain regions (e.g. Great Barrier Reef, wet tropics, south-west corner). Australia has a rich and unique assemblage of vertebrates and vascular plants. By world standards, the number of freshwater fish species in Australia is low, reflecting the small amount of freshwater habitat.

The importance and dominance of invertebrates as a component of biodiversity, and their role in ecosystem functioning is well documented (Ponder & Lunney 1999). The overwhelming number of species and the lack of fundamental taxonomic activity is regarded as being the major impediment to effective invertebrate conservation in Australia and throughout the world. Estimates of total numbers of invertebrate species are inherently unreliable because of the lack of relatively complete taxonomies for the most species-rich taxa. Recher and Majer (1996) (based on a study where they sampled 1600 invertebrate species on four different species of eucalypt) estimated that around 250 000 species of just one group of terrestrial invertebrates would be found in association with the genus *Eucalyptus*.

In contrast, Yen and Butcher (1997) estimated that in total there were around 300 000 species of non-marine invertebrates, of which less than 100 000 are described. Whatever the number, invertebrates make a major contribution to species diversity in Australia (Ponder & Lunney 1999). With the exception of a few charismatic groups, most notably the butterflies, there is very little biological or ecological knowledge of these species.

Microorganisms, are also an important component of species richness in Australia and can have utilitarian and indirect values in both marine and terrestrial systems. Very little is known about these organisms, although estimates have been made of the species richness and endemism of fungi, based largely on their visible fruiting bodies. To illustrate their diversity, May and Simpson (1997) estimated that the approximately 700 species of eucalypts in Australia are likely to have about 7000 species of associated fungi.

Even less is known about other microorganisms. For example, so little is known about bacterial diversity in Australia that even an estimate of the total number is unavailable. However, recent studies are starting to shed some light: Gordon and Fitzgibbon (1999) identified 90 species of bacteria in 20 genera in the gastrointestinal tract of Australian mammals. As with invertebrates, these and other microorganisms are poorly studied and are usually underrepresented in species lists and biological surveys. The number of plant species present at a site can also be underrepresented if surveys of the soil seed bank are not undertaken. For example, a recent study in the wheat belt of Western Australia found that at some sites over 50% of native plant species were detected only as seeds in the soil.

THE FASCINATING WORLD OF MICROBIAL BIODIVERSITY

The microbial world encompasses a large proportion of life forms, broadly including any organism that spends all or most of its life in microscopic form. Microorganisms therefore include viruses, single-celled algae, bacteria, archaea (a specialised form of bacteria that are important in decay processes, especially in wetlands), protozoa and fungi, and in the broadest sense, microscopic invertebrates such as nematodes and mites. During the 1990s, it has become increasingly apparent that our knowledge of biodiversity in the microbial world is largely inadequate. More than 99% of microorganisms are yet to be discovered or described (Amman et al. 1995; Head et al. 1998; Hugenholtz et al. 1998). Understanding the roles of these unknown organisms is likely to be essential to understanding and monitoring biodiversity, since they occupy key positions in all ecosystems. The prokaryotes (bacteria and archaea) as a group contain roughly equivalent quantities of organic carbon as plants, and may hold up to 10 times more phosphorus and nitrogen than do plants (Whitman et al. 1998). Similarly, in terrestrial ecosystems, the biomass of fungi exceeds all other groups except vascular plants (Lal 1995). Microorganisms perform key ecosystem services such as nitrogen fixation, carbon cycling and the regulation of atmospheric gases. A recent attempt to value the world's ecosystem services arrived at a value of US$33 trillion per year (Costanza et al. 1997). Categories of these services that are mainly, or partly provided by microorganisms (nutrient cycling, waste treatment/degradation, atmospheric gas regulation, erosion control, soil formation, biological control and food production) amount to over 70% of this total value.

A further area where we lack adequate understanding of microbial biodiversity is in the associations between microbial life and larger organisms. Almost all macroorganisms depend on microbial symbioses to some extent. Microorganisms contribute significantly to the conservation and production of nutrients in the vertebrate gastrointestinal tract (reviewed in Stevens & Hume 1998). Investigations of the microbiota associated with native Australian animals are only just beginning. Around 75% of vascular plants form mutualistic associations with mycorrhizal fungi. Although the composition and type of plant species in a terrestrial ecosystem is a primary determinant of ecosystem productivity and sustainability (Tilman et al. 1996; Hooper and Vitousek 1997), plant biodiversity may in turn be primarily regulated by the diversity of mycorrhizal fungi (Read 1998; van der Heijden et al. 1998). Consequently, fungal diversity may indirectly control both ecosystem productivity and variability (Read 1998; van der Heijden et al. 1998). Fungal endophytes also occur in the leaves and stems of vascular plants, and the extent and importance of this form of mutualism is receiving more attention. In Australia, 95% of ectomycorrhizal fungi are novel, with some 22 genera and 3 families being endemic (Castellano & Bougher 1994). Thus, Australian fungi are likely to be as unique as our animal and plant species. A similar situation occurs in the mutualistic association between nitrogen-fixing soil bacteria and legumes, which in Australia represent about 10% of plant species. A survey of native shrubby legumes recovered 21 genomic rhizobial species, only one of which corresponded to a known species (Lafay & Burdon 1998).

Because of their large biomass, extraordinary genetic diversity and their central roles in many ecosystem processes, the characterisation of microbial biodiversity should be a consideration in any biodiversity assessment. However, the systematic investigation of microbial diversity in Australia has been the subject of few studies, and methods for rapid assessment of the distribution and abundance of microorganisms are still being developed (Liesack & Stackebrandt 1992; Holmes et al. 2000). The conservation of microbial diversity has not yet received the attention given to larger organisms, and indeed there is only one mention of the word 'microorganism' in the body of the CBD (Davison et al. 1999).

Endemism [BD Indicator 10.6]

A taxon (e.g. a species) is considered endemic to a particular area if it occurs only in that area. The proportion of vertebrate taxa in Australia that are endemic is particularly high compared to other countries, with the richness of vertebrate species being largely as a result of a remarkable variety of reptiles. Because of its size, age and geological and evolutionary isolation, over 80% of mammal, reptile and flowering plant species in Australia are also endemic. The degree of endemism in fungi, molluscs and insects is also estimated to be over 80%, and as such, the Australian continent is recognised as a centre of endemism of global significance (Major 1988). While the level of endemism of many taxonomic groups is still unknown, it is likely to be high because of Australia's evolutionary and geological history.

Since the concept of endemism is tied to particular areas, the identification of centres of endemism is dependent on scale (Crisp et al. 2001). At a global level, high levels of species richness and endemism are consistent across most taxonomic groups in Australia and in most environments. Areas of high endemism also occur within regions in the Australian continent. For example, about 5% of Australia's flora occurs in the Stirling Ranges of south-west Australia, an isolated mountain range with some unique ecological characteristics. The high level of vascular plant endemism in this region is evident also at the global level. In a worldwide study, Myers (1988, 1990) identified 18 relatively small areas that are rich in endemic vascular plant species and that are experiencing relatively rapid rates of habitat modification or loss. The only region in Australia on this list is the south-west of the continent. These 18 sites contain about 50 000 endemic plant species (20% of the world's total) in about 750 000 square kilometres (0.5% of the earth's surface area).

Several different patterns of endemism have been observed within the Australian biota. Cracraft (1991) identified 14 recognisable areas in Australia with a unique assemblage of bird species. Crisp et al. (2001) analysed the distribution patterns of around 8500 vascular plant species. Twelve centres of endemism, which were all near coastal, were identified. The lack of centres of endemism in inland Australia was attributed to the selective extinction of narrow endemics driven by extreme climates during the last glacial maximum.

Major centres of both plant endemism and diversity were also examined by Crisp et al. (2001). The regions that met both these criteria were south-west Australia, the Border Ranges between New South Wales and Queensland; the Wet Tropics near Cairns; Tasmania; and the Iron-McIlwraith Range of eastern Cape York Peninsula. The last centre appears to be more significant than recognised previously, and the Adelaide-Kangaroo Island region, which was also identified as important, has previously been overlooked altogether. It is important to have identified areas where high concentrations of species occur, so that they can be sympathetically managed.

Levels of endemism are also high for marine groups such as macroalgae, with southern Australia being of major significance (Zann 1995). Levels of endemism of other marine groups are described. The 5500 km coastline from south-west Western Australia to the border between New South Wales and Victoria is particularly rich in brown algae and red algae. Of these two groups, around 57 and 75% of species are endemic to southern Australia. Recent studies of cave biota in Australia have also discovered high levels of endemism.

Another measure of the uniqueness of Australia's biodiversity is the high level of variation in species richness among communities, and the variation in commonness and rarity among species, termed mosaic diversity. High values indicate complex landscapes with many environmental gradients, and many species with roughly equal abundance. Using this measure, the diversity of species and terrestrial ecosystems in Australia exceeds that of any other continent.

Vulnerable, endangered, threatened or extinct species [BD Indicator 10.7]

The differences between taxa in the percentages of threatened and extinct species reflect the biases in taxonomic and conservation focus. Vertebrates (especially birds and mammals) and in general vascular plants receive much more attention than do invertebrates, non-vascular plants and fungi. This bias also appears in the records of other countries. It appears that the number of nationally endangered and vulnerable species has increased in several groups over the last seven years. In some instances, the numbers of species in these categories may change over time because there have been changes in the abundance or distribution of species. But in many cases, the changes are due to taxonomic revisions resulting in either the creation or loss of new species.

The status of species may also change based on new information without any underlying change in the number or distribution of individuals or in the processes affecting them. New observations result in a reassessment of area of occupancy, extent of occurrence, population size, threat status, trends in population size or other factors contributing to assessment of conservation status.

Plants

The most important change has been an increase in the number of endangered vascular plants. There were 226 endangered species on the official list in 1993, and 517 in 2001. Most of this is because the work on assessing threatened flora is continuing. Most additions are species that have never been assessed before, including some recently described taxa with very restricted distributions. However, there are some that are facing higher levels of threat than they were in 1993. The most important cause of changes in threat status that are not the result of new work or taxonomic revision over the last five years has been land clearance for urban and agricultural development.

As many as 123 species vascular plant species were presumed extinct in 1988 but the number was revised downwards subsequently, largely as a result of new information, survey work and taxonomic revision. A total of 227 vascular plant species have been considered at some time to be extinct. The first two Rare or Threatened Australian Plants (ROTAP database) lists share relatively few species with one another, or with any of the subsequent lists. The turnover in species on the list of presumed extinct plants has been substantial, varying by 10 to 40% of their constituent species, even during the 1990s when the total number of species presumed to be extinct did not change greatly. Most of the changes are due to taxonomic revision and survey work. Few reflect actual changes in status.

The number of presumed extinct vascular plants at the national level has fallen from 74 to 63 during 1993 to 2001. This has occurred as a result of taxonomic revision and rediscovery. Additional survey work is often prompted by the inclusion of species on the 'presumed extinct' list. There may be some further reduction over the next five years. There may be new additions as new species are described, data on plant abundances are revised, and surveys are conducted for rare and endangered flora.

Birds

The 2000 Action Plan for Australian Birds (Garnett & Crowley 2000) listed 25 bird taxa (reporting to the subspecies level) as extinct, 32 as critically endangered, 41 as endangered, 82 as vulnerable and 81 as near threatened. The remaining 1114 taxa are deemed to be least concern, including 28 introduced taxa and 95 vagrants. Of those taxa known to have been present or to have occurred regularly in Australia when Europeans settled in 1788, 1.9% are reported as extinct and a further 11.5% are considered threatened. Some 6.0% are near threatened. Since the last Action Plan in 1992, research and surveys have shown that seven taxa are less threatened than was thought but a further 56 taxa should have been listed. Other differences between the 1992 Action Plan and Garnett and Crowley (2000) are accounted for by changes to taxonomy (19 taxa), to more rigorous IUCN criteria, which better define the different categories (138 taxa) or both (11 taxa).

Using current knowledge, taxonomy and IUCN criteria, there has been a change in the status of 25 bird taxa (2.0%) over the eight years since the 1992 Bird Action Plan. For seven taxa, the conservation status can be downgraded as a result of effective conservation management: two from critically endangered to endangered, four from endangered to vulnerable and one from vulnerable to near threatened. However, the status of 18 taxa should be upgraded. Although no taxon has become extinct in the last decade, there has been a net increase of eight critically endangered taxa, with six fewer vulnerable and one more near threatened species.

Invertebrates

There are 281 species of Australian invertebrates listed on the 1996 IUCN Red List of Threatened Animals (IUCN 1996), representing fewer than 0.5% of known taxa. However, only four species are listed nationally in the EPBC Act. The first attempt to appraise an invertebrate group in Australia comprehensively is underway for butterflies. At a broader level, the draft Action Plan for Invertebrates aims to objectively assess the conservation status of invertebrates in Australia by using 25 species to illustrate the range of needs for this diverse group.

The conservation status of marine species

Knowledge of the conservation status of most of Australia's marine species is very limited. Australia's first endangered marine fish, the Spotted Handfish (*Brachionichthys hirsutus*), is endemic to the lower Derwent River estuary in Tasmania. The decline in Spotted Handfish numbers has been linked to decline in suitable spawning substrate due to overall deterioration in the health of the Derwent system and the effect of the introduced Northern Pacific Seastar (*Asterias amurensis*).

Scientific interest has largely centred on the higher vertebrates such as turtles, seabirds, seals, Dugongs and whales. Microorganisms, algae, invertebrates and fish have been generally neglected. Australia is very rich in macroalgae or seaweeds. Southern Australia has over 1150 species (Zann 1995). This is greater than 50% more than any comparable region in the world.

Fish Species

Australia has an estimated 4000 to 4500 species of fish, of which around 3600 have been described (Zann 1995). About one-quarter of the species are endemic, most of which are found in the south. Although regulations governing many of the fished species have long existed in Australia, marine fish conservation is a relatively new field and the conservation status of most species is poorly known. Potentially vulnerable fish include sharks, which are slow growing, have a low reproduction rate, are highly migratory and form schools during the mating season. Threats are commercial and sports fisheries, and shark meshing of surfing beaches. Fish species with restricted distributions are also vulnerable, particularly from loss of habitat. Broad-scale studies of the distribution of coral reef fishes, conducted by the Australian Institute of Marine Science (AIMS), show that overfishing can rapidly deplete the stocks of coral reef fishes. Surveys of the density of Coral Trout on Bramble Reef during and after the reef was closed to fishing found that the population of legal size Coral Trout fell by 57% in just two months when the reef was reopened for fishing (Wachenfeld et al. 1998).

Sea Snakes

Australia has about 30 of the total number of 50 species, about half of which are endemic (Zann 1995). The family of aipysurids live in coral reef waters and the family of hydrophiids live in the interreef waters of Australia's tropics. Sea snakes bear live young and have a relatively short lifespan; they reach sexual maturity in around three years, and live for some 10 years. The greatest human impact is from prawn trawling. Between 10 and 40% of sea snakes taken in trawls die once released (Barratt et al. 2001). For the past 20 years, trawled sea snakes have been used in a small leather industry. Licences limit the take of sea snakes for leather to 20 000 per year (Barratt et al. 2001).

Turtles

Unlike sea snakes, turtles are a long-lived group of reptiles that are slow to reach maturity. They may breed only around five times in their lives, making them extremely vulnerable to overexploitation and habitat destruction or modification (Zann 1995). Breeding migrations may cover hundreds to thousands of kilometres and many turtles breeding in Australia may live around the islands of Papua New Guinea, the south-west Pacific Islands and Indonesia, making habitat management and enforcement difficult. The main human effects that occur while turtles are in Australian waters are: mortality of subadults and adults in prawn trawls, shark nets, drumlines and gill nets, and in collisions with high speed vessels; hunting by Indigenous communities; habitat degradation; and predation on eggs by feral animals. The effects of disease and parasites are unknown.

Large dots denotes thousands of nesting females per year, medium denotes 10 to 100 females per year and the smallest dots are <10 per year. There is a gap in knowledge for Arnhem Land and data for Western Australia has been pooled at the regional level.

Monitoring of turtles is essential to ensure that management practices are suitable to minimise or eliminate habitat degradation. In the Great Barrier Reef World Heritage Area most scientific studies of turtle populations have concentrated on Green and Loggerhead Turtles (*Chelonia mydes and Caretta caretta* respectively) (Wachenfeld et al. 1998).

The Loggerhead is of specific concern as the number of nesting females has steadily declined since surveys began in the late 1970s. The east Australian population of Loggerhead Turtles used to represent the bulk of the South Pacific stock. If this population disappears, it will represent a highly significant loss (Wachenfeld et al. 1998). Because female turtles nest in the area where they were hatched, it is highly unlikely that a stock that has died out would be colonised naturally by Loggerhead Turtles elsewhere in the world.

Seabirds

About 142 species of seabirds belonging to 12 families are found in Australia and its external territories (Zann 1995). Of these, 76 species breed and spend their entire lives in the region, and 34 species are regular or occasional visitors. Problems for sea birds include illegal poaching of adults, chicks and eggs; mortality from bushfires and feral animals; incidental capture of albatrosses and other seabirds by longline fishing; clearing of habitats; decline in prey due to overfishing; and disturbances of nesting colonies by humans and low-flying aircraft (e.g. Wachenfeld et al. 1998; Barratt et al. 2001). Possibly half of Australia's nesting islands are subject to one or more of these direct human threats.

Dugongs

The tropical Dugong (*Dugong dugon*) is the only fully herbivorous marine mammal and the only Sirenian (sea cow) to occur in Australia (Zann 1995). It is extinct or near extinct in most of its former range which extended from East Africa to South-East Asia and the Western Pacific. Northern Australia has the last significant populations (estimated to be over 80 000) in the world. Large, long-lived mammals, Dugongs become sexually mature at around nine to 17 years and calve every three to seven years, making them vulnerable to excessive mortality. Management concerns include the potential for overhunting of some Torres Strait populations, death of individuals that are accidentally caught in fish gillnets and shark nets, and loss of seagrass habitat (Zann 1995; Wachenfeld et al. 1998).

Surveys of Dugongs have been undertaken across Queensland since the early 1980s. Surveys south of Cooktown have documented a distinct decline in population with the 1994 estimate being only 48% of that for 1986 to 1987. A major mortality of animals occurred in Hervey Bay (Qld) in 1992 following the die-off of seagrasses. The Dugong is listed by the IUCN as 'vulnerable to extinction' and is a listed marine species under the EPBC Act. A population of around 10 000 animals at Shark Bay in Western Australia is considered of major significance for the species.

Seals

Australia's seals were overhunted in the 1800s. They are now fully protected and some populations appear to be increasing, although marked long-term declines have been reported for other populations such as the Southern Elephant Seal (*Mirounga leonina*) on Macquarie Island which have been monitored for the last 50 years. In this instance, the cause of the decrease is unknown, although it is likely to be related to increased competition for food supplies.

An Action Plan has also been developed to encourage the long-term viability of seals. Major human threats include entanglement in fishing nets and ocean litter, oil pollution and disturbances by visitors. Fur seals are still occasionally illegally killed for lobster bait, and around fish farms for 'stealing' fish. Development of predator-resistant cages has reduced the latter problem. Entanglement in nets and plastic box straps remains a major threat. About 2% of seals at haul out or resting sites in Tasmania are entangled in net fragments and other plastic litter at any time (Barratt et al. 2001). A significant number of more badly tangled seals drown before reaching haul out sites. In 1990, an oil spill in Western Australia affected a number of New Zealand Fur Seal pups (Zann 1995).

Whales and Dolphins

Gillnets, shark nets set off bathing beaches, discarded fishing nets, bioaccumulation of toxins and ingestion of plastic litter are considered threats to cetaceans (whales and dolphins) within Australia (Zann 1995). During the 1980s, almost 14 000 dolphins were drowned in Taiwanese shark gillnets off northern Australia but this fishery is now closed. The use of long driftnets (sometimes referred to as the 'walls of death'), which caused substantial mortalities of cetaceans, is now banned under the Convention for the Prohibition of Fishing with Long Driftnets in the South Pacific and the United Nations global moratorium on their use. However, many cetaceans are still caught in protective shark nets off bathing beaches. For example, around 520 dolphins were caught in shark nets off the coast of Queensland between 1967 and 1988 (Zann 1995).

During the period of industrial whaling (1948-1962), the estimated numbers of Humpback Whales (*Megaptera novaeangliae*) fell dramatically. Whaling has been replaced by the new industry of whale watching, which can bring significant incomes to local economies. For example, in Victoria, Warrnambool's land-based whale-watching industry is estimated to contribute $17 million annually to the region. Because of concerns that boats, aircraft and divers may affect whale behaviour, regulations govern the distances that observers may approach whales. Increases in the estimated number of whale species such as the Humpback Whale in eastern Australian waters suggest that habitat conditions are sound. Numbers have shown a steady increase since regular monitoring began in 1981.

Recovery plans: Securing species and communities in the wild

Recovery plans set out the actions thought necessary to support the recovery of threatened species or ecological communities to maximise their chances of long-term survival in the wild.

There were 1478 species and 27 ecological communities listed under the EPBC Act in February 2001, as either endangered or vulnerable at the national level. The categories 'critically endangered', 'conservation dependant' and 'extinct in the wild' have been added to the previous categories of endangered, vulnerable and extinct for threatened species and 'critically endangered' and 'vulnerable' have been added to the previous category of endangered for ecological communities. When approved by government Ministers, recovery plans become statutory documents. Furthermore, the EPBC Act prohibits Commonwealth agencies from undertaking activities that are inconsistent with a recovery plan.

As of May 2000, there were 37 adopted recovery plans in place under the EPBC Act covering 44 species, 18 of which are plants. This means that around 3% of nationally listed species and communities have recovery plans. In early 2001, another 100 recovery plans, covering in excess of 130 species were being considered for adoption by the Threatened Species Scientific Committee. State and territory governments also prepare recovery plans under their legislation, but documenting these plans was beyond the scope of this report.

There are many more plans in preparation, many without any funding support from the ESP. The amount of funding provided for the implementation of recovery plans, compared with the amount specified in the plans for full implementation. This information relates to recovery plans funded through the ESP for species listed in the (now repealed) *Endangered Species Act 1992*. The total cost of implementation of the adopted plans from all sponsors comes to $25 015 105, of which the ESP provided $8 832 275, about 35% of the total cost. In order to be recognised under the EPBC Act, recovery plans have to be resubmitted.

Looking after species away from their homes

As part of the recovery process for threatened species, zoos and herbaria have been refining the processes through which they manage captive populations. Although managing species in the wild is preferred, sometimes it is necessary to manage them *ex situ* (away from their homes). Usually this is required to increase the number of individuals by breeding them in captivity and releasing them into the wild when conditions are suitable. Considerable human and financial resources have been expended on *ex situ* conservation for threatened species.

MALA RECOVERY PROGRAMS AND MALA DREAMING

The Mala or Rufous Hare-Wallaby (*Lagorchestes hirsutus*) once occupied about 25% of the continent and was common across the spinifex deserts of the Northern Territory and north-west South Australia, to the temperate woodlands and grasslands of Western Australia's wheat belt. The species was an important food source to Indigenous peoples throughout its geographical range, and the Mala remains of great cultural significance to the people of the central deserts. Since European settlement, the population of Mala has declined to the brink of extinction, suffering a similar fate as many small to medium-sized mammals in arid Australia. It is now listed as a nationally endangered species.

By 1980, there were only two known populations of the species on the continent, comprising a total of about 50 individuals. These populations were situated close together in the Tanami Desert in country owned by the Warlpiri. In 1980, a captive breeding program for the species was commenced in Alice Springs. By 1986, as a consequence of successful breeding, scientists from the Northern Territory's PWC were able to translocate progeny of this population into an electric-fenced enclosure on the floodplain of the Lander River, near the Indigenous community at Willowra in the Tanami Desert.

Despite numerous efforts to reintroduce Mala from the enclosure into adjacent spinifex grassland, a wild population could not be established mainly because of predation by the cat and fox. The captive breeding program at the Alice Springs Desert Park continues under the direction of the PWC. Translocations of some animals bred in central Australia have been made to habitats previously occupied by the species in Western Australia, including an enclosure in Dryandra forest south-east of Perth, and to breeding compounds in Franois Peron National Park. Another group of Mala have been successfully established in the wild on Trimouille Island off the Pilbara coast. Recently Mala from the Tanami Desert captive colony were translocated to a large new enclosure in Watarrka National Park to the south-west of Alice Springs.

The Mala occupies an important role in the cultural lives of the Warlpiri and other Indigenous peoples who are traditional owners of the land the species once occupied. The Mala features prominently in the creation stories and ceremonies of many of these peoples, and the Warlpiri felt a considerable responsibility for the disappearance of the species as they relate its decline to a decline in traditional Mala ceremonies.

Mala (Wallaby) Dreaming was the first painting by Kumantjayi Tjupurrula, who is now recognised as one of Australia's greatest contemporary artists. The dreaming or story of the painting was recorded by Geoffrey Bardon AO, when it was created at Papunya (NT) in early 1971. He noted (G Bardon 1999, *Summary of story associated with painting*):

The central pattern of the work shows ceremonial men sitting at a series of fireplaces. On both sides are wallaby tracks from the Wallaby Spirit Men as they travel to and from a sit-down place, which are indicated by the concentric circles. The repetition and looping between concentric circles in the central motif, represents a dancing and singing ceremony. This ceremony is held in turn at a sit-down place that is part of the Wallaby Dreaming.

Craig et al. (1998) provided some material on the release of captive-bred animals from zoo programs. Of the 19 programs that explicitly recognised the process by which zoos cooperate with wildlife agencies as part of an overall recovery plan, captive breeding for release into the wild is an explicit goal of 16 recovery programs. The recovery program for the Mala, an endangered species with particular importance for Indigenous people in central Australia illustrates the complex nature of these activities. Predation by feral cats and foxes pose a major threat to species when they are released into the wild and major programs such as the Western Shield aim to reduce their numbers before native species are released into the wild.

Species of economic importance

The values of biodiversity discussed at the beginning of this report included economic values. The economic value of some native species or suites of species is well enough known, as there are recognised economic activities and products traded in markets. These include major industries such as fisheries and forestry, which are based largely on wild populations. The wildflower, bushfood and cut foliage industries involve numerous native species. Perhaps less obvious is the use of spiders and snakes for their venom and the harvesting of sea cucumbers - an animal that is considered a delicacy in Asian countries. Seahorses are also harvested from the wild, being primarily used as Chinese medicines, aphrodisiacs and food, although the market for curios and aquarium fish is growing steadily. Tropical species of seahorse are endangered as a result of wild harvesting. A company in Tasmania is planning to supply this market by farming temperate species. The species used is proving suitable for aquaculture and the venture is on the point of commercial production.

Any attempt to assess the economic significance of biodiversity is complicated by two factors. First, many values of biodiversity are not traded in markets. Thus, new and often contested valuation methods must be used to estimate the financial value of biodiversity. This difficulty is encapsulated by the example of a valuation of use of an Australian national park, where one critic claimed the original estimate was four times too large and another claimed it was six times too small (Beal 1998).

Second, many biodiversity-based industries are recent and small and thus often do not have well-developed information systems. Third, biodiversity is the classic 'cross-sectoral' issue, so that expenditure or economic activity across a wide range of commercial sectors and policy portfolios affects or is affected by biodiversity.

ASSIGNING A DOLLAR VALUE TO BIODIVERSITY

Although there is considerable and increasing activity that gains economic benefits from native species, few empirical data are available. Consequently, the following examples are indicative only.

Commercial fisheries

Commercial fisheries in Australia depend largely on wild caught species, supported an estimated production value of $1.8 billion in 1998-99 (ABARE 1999).

The oyster industry, based on the Giant Australian Oyster (*Pinctada maxima*) produces exports of some $200 million per year. There is major scope for commercial production of some native fish species for human consumption, such as Silver Perch (*Bidyanus bidyanus*), Murray Cod (*Maccullochella peelii*), Snapper (*Chrysophrys auratus*), Barramundi (*Lates calcarifer*) and Mulloway (*Johnius antarctica*).

Freshwater eels (*Anguilla* spp.), which have a poorly understood breeding migration to offshore waters (possibly in the Coral Sea), support an industry of about $5 million per year and are increasingly 'ranched', with harvested young raised in captivity.

It is likely that the total economic value of recreational fisheries, again largely based on native species, exceeds that of commercial fisheries when ancillary activity (e.g. boats, fuel, accommodation) is accounted for.

Forests and Fodder

Woodchips from native forests had a value in 1998-99 of $590 million. While this is only one category of forestry production in economic terms, it is a particularly significant one and known to be sourced from native forests rather than exotic plantations. Significant pastoral industries in drier areas in Australia depend on pasture provided by native shrub, grass and forb species. For example, 7% of Australia's beef cattle are found in the Northern Territory, and these depend largely on native pastures. To give some estimate of this value, between 1983 and 1984, the value of native pasture support industries in Queensland was calculated as $1125 million per year. There are increasing efforts to re-establish useful native shrub species for fodder and land degradation control (e.g. Old Man Saltbush, *Atriplex nummularia*, in western New South Wales).

GROWING AND HARVESTING NATIVE PLANT SPECIES

The ABS (ABS 1999a) used the annual service costs of maintaining a garden with native plants as the basis for estimating households' expenditure towards 'protection of biodiversity and landscape'. The result was an estimate of $169 million in 1996-97, which reflects a growing use of native species for ornamental purposes. Australian wildflower exports in 1997 were valued at $30 million and the 'bushfood' industry was expected to grow from $14 million in 1996 to over $100 million in 2000. There are no data on the value of the native species component of the nursery industry, but some 1600 species are grown and traded (Parliament of Victoria 2000).

Native Animal Species (meat, skin and hides)

Duck and quail hunting (recreational) in Victoria is estimated to involve over $30 million expenditure each year. The kangaroo industry is estimated to support 4000 jobs in rural areas, and the value of the industry is about $245 million per year; the emerging Emu (*Dromaius novaehollandiae*) industry is smaller at $6-8 million per year. In Tasmania, the Brushtail Possum (*Trichosurus vulpecula*) harvest has an annual value of $400 000, the Muttonbird (*Puffinus* spp.) harvest $425 000 and Bennett's Wallaby (*Macropus rufogriseus*), $750 000 (DPIWE 1999). An increasing number of possums are exported for human consumption in China as a 'warming meat'.

In many such industries, there is considered to be scope for greater economic value through increased processing, niche products, expansion of cultivation rather than harvesting, or greater use for human consumption (e.g. much kangaroo meat is used for pet food, and much leather remains unused). Again, different values will define different attitudes to the expansion of such industries. Some biodiversity-based industries now coordinated by industry organisations are more organised towards both sustainable practices and value-adding and export. These include the Kangaroo Industry Association and the Southern Bushfoods Association.

Live Animal Species

Another indication of the economic value of Australian species is the prices paid for single specimens for companion animals. Red-tailed Black Cockatoos (*Calyptorhynchus magnificus*), one of the few bird species subject to a managed harvest for export (from NT), can fetch $8000 to $9000 per animal, the same as a Gang-Gang Cockatoo (*Callocephalon fimbriatum*). Sulfur-crested Cockatoos (*Cacatua galerita*) and Galahs (*Cacatua roseicapilla*) can fetch over $1000 each, as can a native Green Tree Python (*Chondropython viridis*) or a Spiny-tailed Monitor (*Varanus acanthurus*) in other countries. This trade is small, and beset with problems of legality and whether this is an acceptable or appropriate use of biodiversity.

Some argue that if wild populations have a commercial value, landholders are encouraged to improve the resources they depend on and that a carefully regulated industry could increase the distribution and numbers of wild species. Earth Sanctuaries Limited has put an 'economic value' on a range of vertebrate species and includes $3.8 million for Australian fauna on its balance sheet. For example, the 130 individuals of Numbat owned and managed by the company are valued at $650 000. The Productivity Commission has recently undertaken a study of creating markets for biodiversity resources and services that focuses on Earth Sanctuaries.

In Western Australia, Fund for Wild Australia, a privately owned non-profit organisation, is also developing a series of wildlife exclosures to protect endangered mammal species, although this enterprise is not listed on the stock exchange. The World Wildlife Fund for Nature argues that while well-managed predator exclosures can contribute to the conservation of certain species, their benefits should not be exaggerated. They should be seen as just one of the tools available to help protect endangered animal species.

Tourism

Other species are clearly of economic significance but this is more difficult to estimate. Tourism provides an example. Hundloe and Hamilton (1997) estimated that the koala is worth $1.1 billion per year through its iconic role in attracting international tourists to Australia, and that in the absence of Australia's unique wildlife, there would have been a loss of $1.8 billion in tourism revenue in 1996 (rising to $2.5 billion in 2000). Surveys of international tourists indicate that a major proportion identify nature-based factors such as wildlife and national parks as a motivation for their visit. Many remote communities are increasingly dependent on income from small-scale 'ecotourism' ventures.

Although the contribution of particular elements of biodiversity to the attractiveness or allure of, say, a national park is difficult to measure precisely, it is fundamental. Valuations reported in Bennett et al. (1996) provided some idea of the local and regional benefits of protected areas. The annual economic value of Dorrigo and Gibraltar Range National Parks in New South Wales was $5.4 million and $800 000, respectively. Such figures and associated employment can be highly significant in rural areas. Biodiversity in protected areas is obviously crucial to the tourism industry, both domestic and international. Some indicative economic values from Victoria are: nature-based tourism in the Grampians region, $100 million annually; and the Penguin Parade Reserve, $96.5 million in 1995-96 and over

1000 jobs (Parliament of Victoria 2000). Visitation to national parks and reserves in Victoria grew from 8 to 25 million per year between the late 1980s and late 1990s.

Bioprospecting

Bioprospecting (the chemical prospecting for pharmaceuticals in natural organisms) is a growing industry in Australia, with potential in both terrestrial and aquatic environments. If managed appropriately, bioprospecting has the potential to have minimal impact, as modern screening methodologies and analytical instruments permit the identification of biologically active compounds from fairly small samples (Benkendorff 2001). Another area of clear future significance is the potential economic value of native genetic resources. The ability to capture new biotechnological benefits will rely on maintaining biodiversity in its natural environment because the exploitation of metabolites usually depends on observing the interactions between organisms in their natural environment (Battershill & EvansIllidge 2000).

BIOPROSPECTING: CHEMICAL PROSPECTING IN NATURAL ORGANISMS

Bioprospecting - the chemical prospecting for pharmaceuticals in natural organisms - has been promoted as a means for discovering new medicines, an instrument for economic development and an incentive for conservation. Certainly most of both the drugs in commercial use, and those being developed, are of natural origin. The most well-known and celebrated example of a pharmaceutical company channelling money towards the preservation of biodiversity is the collaboration between Merck and the Costa Rica National Biodiversity Institute (INBio). Merck negotiated an up-front fee of US$1 million for the opportunity to explore Costa Rica's biodiversity for novel drugs. If a useful drug is discovered, INBio will receive a share of the royalties. As part of the deal, 10% of the up-front money and 50% of the royalties must go directly towards conservation.

A major program at AIMS is examining the benefits from marine biotechnology. The aim is to use specimens of macro and microorganisms to discover novel biomolecules with strong biocidal and anti-infective activities as well as developing and commercialising technology for seafood diagnostic kits.

Benkendorff (2001) identifies conservation benefits and problems associated with bioprospecting in the marine environment. Comprehensive surveys undertaken as part of her studies recorded a much higher species diversity than previously recognised, identified an important breeding site and hotspot of molluscan species richness, recorded new distributions for three species and found an unidentified Polycerid nudibranch. Following the surveys, her research focused on the marine mollusc, *Dicathais orbita*. Several potential resources are produced by both the adults and egg masses of *Dicathais orbita*, including pharmaceuticals.

Several ethical issues have also been identified such as the potential environmental effects of extraction, the need for the fair and equitable sharing of results and benefits, and the need to protect intellectual property rights when traditional or other knowledge about the natural biota is shared with bioprospectors.

Benefit sharing and intellectual property rights, and to a lesser extent the environmental effects of bioprospecting, have been described in several national and international declarations, resolutions and other publications. These include the recent public inquiry into access to biological resources in Commonwealth areas (Voumard 2000). The final report from this inquiry recommended that bioprospecting should be considered as a 'matter of significance' under s23 and s26 of the EPBC Act when assessing applications for access to biodiversity on Commonwealth land or in Commonwealth marine areas. State governments are also developing policies on the access and benefit sharing of biodiversity resources. The Commonwealth House of Representatives Standing Committee on Primary Industries and Regional Services is also examining the development of high technology industries in regional Australia based on bioprospecting. At the international level, bioprospecting is a major topic being examined through the CBD, with Australia being strongly involved in the development of a policy on access and benefit sharing. Given this level of activity, it is likely that the level of bioprospecting in Australia will increase, despite the financial risks and potential difficulties associated with synthesising large quantities of the extracted compounds.

Ecosystem Services

Such estimates of particular economic contributions from biodiversity only go some way towards providing measurement of the total value of biodiversity. For example, although the $300 million-plus Australian honey industry is based on introduced bees, these bees depend greatly on native plant species for pollen and nectar. The value to agriculture of 'pollination services' by native insects as well as honeybees is likely to be worth significantly more than this.

The actual economic and social value of 'ecosystem services' (indirect utilitarian values) is often difficult to calculate. For example, the role of soil organisms in maintaining agricultural production is both poorly known and of obvious economic significance. However, the service values of ecosystems can be highlighted when they become degraded to the point that the economics of restoration are measured in dollar terms. In the mid-1990s the treatment of land degradation in Australia, for example, had direct costs of more than $400 million per year, including treatment for waterlogging, salinity and erosion. More recent estimates by the Australian Conservation Foundation (ACF) and National Farmers Federation have put the annual cost of degradation in rural landscapes at a minimum of $2 billion, which is predicted to increase to over $6 billion annually by 2020 if no action is taken (Madden et al. 2000). The recently released report *Coordinating Catchment Management* (from the bipartisan House of Representatives Standing Committee on Environment and Heritage) recommended that a National Environment Levy be put in place for the next 25 years to help fund programs to address these issues. Carbon and biodiversity credits, while still in their infancy, propose to calculate specific dollar values on elements of biodiversity so they can be traded on domestic and international markets. Interest has increased in valuation of non-traded ecological services and assets, especially in the wake of a global study that estimated the economic value of 17 ecosystem services across 16 biomes as between US$16 to 54 trillion ($10^{12}$) per year, with an average of US$33 trillion per year (Costanza et al. 1997). This valuation was, in turn, used to estimate an average value in 1997 for terrestrial Australian ecosystems of US$245 billion per year and for marine ecosystems of US$640 billion per year (Jones & Pittock 1997). Whether such valuations will, first, develop widely accepted methodologies and, second, ever become used routinely in decision making and biodiversity policy, will only become clear with time. It may be that the main use of broad ecosystem service valuations will influence public perceptions rather than, say, feeding into cost-benefit analysis at a project approval level. In Australia, the CSIRO, supported by the Myer Foundation, has commenced a major research project into the valuation of ecosystem services.

Some native species may be considered to have negative economic value, as a result of reduction in productive capacity of the land. An example is the increase in 'woody weeds' in central and western New South Wales and Queensland (e.g. *Senna* and *Eremophila* spp.), although in such cases there may be counterbalancing benefits such as protection of wildlife habitat and soil.

THE USE OF NATIVE SPECIES BY INDIGENOUS COMMUNITIES

Some small-scale biodiversity-based industries are significant to small or remote settlements and to Indigenous communities (Australian Senate 1998). The kangaroo industry is one example, as is the Tasmanian Muttonbird (*Puffinus tenuirostris*) harvest that is carried out largely by Tasmanian Indigenous peoples and is considered to be sustainable. Crocodile 'ranching', where wild young are harvested and then grown in captivity is important to some Indigenous communities in northern Australia, and crocodile 'cruises' for tourists was generating some $2 million of economic activity per year in the mid-1990s. Increased breeding and stocking in recent decades of native fish species, especially Golden Perch (*Macuaria ambigua*) and Murray Cod (*Maccullochella peelii*), has established several inland empoundment fisheries of considerable local economic importance in mainland eastern Australia. The genetic implications for wild populations of the widespread distribution of these stocks may be an issue.

Access to biological resources has been the subject of a recent major public inquiry (Voumard 2000). This inquiry gave advice on an access scheme that could be implemented under the EPBC Act. The scheme centred on a benefit-sharing contract which included protection for and valuing of Indigenous knowledge and environmental benefits in the areas from which the resource was obtained. Although the inquiry focused on Commonwealth areas, it recommended that a nationally consistent scheme be developed across all jurisdictions.

Species known to be changing significantly in distribution

The Australian environment has undergone, and is still undergoing, significant and often detrimental changes. Threatening processes such as broad-scale land clearance, overharvesting and altered fire regimes would be expected to significantly change the distribution of native species. For example, loss of habitat in southern Australia is associated with a reduction in the range of the Magpie Goose (*Anseranas semipalmata*).

At a national scale, the most comprehensive database available on the distribution and relative abundance of native species is on birds (Barrett 2000b). Between 1977 and 1981, 90 000 bird surveys, collected by 3000 volunteers, were entered into computer by hand - and thus the first continental-scale Bird Atlas (referred to as the Field Atlas) was created. Saunders et al. (1998) recommended that a repeat survey be initiated to reflect changes over 20 years. This is underway, and like the Field Atlas, the current version (referred to as the New Atlas) is an Australia-wide, community-based survey being conducted over several years.

The two data sets were used to analyse whether there had been significant changes in the distribution of bird species at the national level over 20 years. Although the New Atlas survey is yet to be completed, preliminary analysis has shown changes in the distribution of bird species in southern and eastern Australia. The analysis was confined to these areas as differences in sampling meant that northern Australia has been undersampled in the second Atlas, at least at this point. Other biases include sampling along roads in the more remote parts of Australia and a bias towards recording common open country birds in the Field Atlas due to differences in sampling techniques used.

The analysis detected a total of 65 species that appear to display reductions in range between the Field Atlas (1997-81) and the New Atlas (1998-2000). These differences are sufficiently large that they are likely to be biologically important. Of these species, 13 show a substantial and systematic difference between the Field Atlas and the New Atlas surveys. The list includes birds of prey, ground birds and water birds, illustrating that a range of changes could be responsible for the major reductions in range.

To try to understand why these 13 species appeared to show major declines in range in the last 20 years, opinions were sought from five specialists from around Australia.

There was consensus among four specialists that at the national scale the following species had genuinely contracted in their range due to human-related threats: the Brush-turkey (habitat loss and degradation/predation), Australian Bustard (habitat loss and degradation/hunting), Black-chinned Honeyeater (loss of habitat, especially mature trees), Fuscous Honeyeater (loss of habitat) and the Wedge-tailed Eagle (loss of food, rabbits, due to the calicivirus). Changes in the range of the Black Kite were thought to be related to differences in rainfall between the two Atlas periods. Below average rainfall was recorded over the period of the first survey.

During such dry climatic periods, there can be a considerable expansion in the range of bird species in the search for water and prey.

Three specialists commented on species in southern Australia. Of these, two thought that the decline of the European Goldfinch could be related to the intensification of agriculture and the decline in preferred weed seeds. Changes in agricultural practice were also associated with the Australian Pipit, although one commentator felt that it might be returning to its pre-European range. The raw data for the White-throated Needletail, with the range analysis, picked up a significant decline. One specialist felt that this was an artefact of the differences in sampling between the two Bird Atlases, another that the decline was related to pressures on its habitat in South-East Asia and one-third gave no explanation for the observed pattern. These responses demonstrate that even when there appears to be a major decline in the range of a species, informed opinion about the reasons underlying the change can vary.

Some introduced species such as the Common Myna (*Arcdotheris trisis*), Common Blackbird (*Turdus merula*), Mallard (*Anas platyrhynchos*) and Spotted Turtle-dove (*Streptopelia chinensis*) as well as native species able to exploit urban environments such as the Eastern Rosella (*Platycercus eximius*) and Red Wattlebird (*Anthochaera carunculata*) appear to have expanded their range since 1981. While the results on the species that are increasing could be an artefact of the analysis, other sources of information support the results.

As noted earlier, the analysis undertaken for this report of the Bird Atlas data focused on species in southern Australia. In a separate study, Franklin (1999) used historical data on the distribution of birds to assess change in groups of grain-eating birds in the tropical and subtropical savannas of northern Australia. Twelve of the 49 native and mostly resident species had declined, and three others had increased. One species was thought to be extinct (the Paradise Parrot, *Psephotus pulcherrimus*), and two taxa were critically endangered. Even though human settlements are sparse in these areas and the intensity of pastoralism is relatively low, these changes demonstrate the potential effect of altered fire and grazing regimes in these northern woodlands.

Information is becoming increasingly available on changes in distribution of plants and animals at a local and regional level. A relevant example is a recent study on reptiles in central-western New South Wales that showed that two species have vanished from the agricultural landscape. Up to a further nine species were in decline, and may be facing the same fate. Extensive land clearing appears to be the reason for the reduction in the range of these reptiles, with inadequate native vegetation remaining to satisfy their need for food and shelter.

Analysing changes in bird distribution using data from the Field Atlas (1977-1981) and the New Atlas (1998-2000)

Bird atlas data were supplied by Birds Australia to analyse changes in the distribution of bird species over 20 years. Two sets of Atlas data were supplied; the first Field Atlas spanned from 26 December 1972 to 30 June 1986, but data used in this analysis were restricted to 1 January 1977 and 31 December 1981. The data set was cleaned of obvious errors. For example, 313 survey sheets plotting at least 30 km offshore were deleted. A total of 87 595 sheets observed at 12 200 unique locations were accepted resulting in 2 649 449 species observations.

The New Atlas data span the period from 1 January 1998 to 3 July 2000 (data collection is ongoing). A total of 138 surveys plotted more than 30 km offshore of Australia were deleted from the database. Taxonomic variations in the New Atlas were corrected to be consistent with the Field Atlas. Species groupings and species not recognised within Australia were removed from the database. A total of 91 983 New Atlas survey sheets remained, at 50 350 unique locations recording 1 688 204 species observations. The relatively high number of unique locations reflects the higher precision (seconds instead of minutes) of most of the coordinates of the New Atlas surveys. Geographical Positioning System (GPS) was used to position about 60% of New Atlas surveys, whereas none of the Field Atlas records were GPS-based. The distinct sampling patterns along roads in central and northern Australia can be detected in the figures from both Atlases.

Taxonomic revisions result in some confounding of data between the two observation periods. Species split into two or more species were grouped under a single atlas number. Similarly, subspecies in the Field Atlas more recently revised to full species status were relegated to their original atlas number because the original records were not resolved. There were sufficient data from the two survey periods to include 492 bird species in the analysis out of the approximate total of 750 species recorded. The subset of species was located principally in southern and eastern Australia because areas in northern and inland Australia had considerably lower numbers of surveys in the New Atlas compared with the original Field Atlas. The most likely cause is because at the time of the analysis, data were available only for 2.5 years for the New Atlas compared with five years for the first Atlas. Consequently, comparing species in these undersampled areas would result in apparent changes in distribution that were an artefact of the data collection.

Perhaps the most impressive migrants that journey to Australia are the shore birds, which are found in shallow water in both coastal and inland wetlands. These birds, which include sandpipers, curlew, snipe and plovers, may fly 2500 km annually as they migrate between the southern and northern hemispheres. Of the 50 species of wading or shore birds that regularly occur in Australia, 33 breed outside Australia in central Asia, Siberia or the Arctic zone of North America (Blakers et al. 1984). One of the important routes that is travelled by these birds is the East Asian-Australasian Flyway, which extends from the Arctic Circle through SouthEast Asia to Australia and New Zealand. This Flyway has a series of wetlands that the shorebirds visit to rest and feed before they undertake the next stage of the journey. When other bird groups are also considered, over 300 species are known to migrate between Australia and other countries (Birds Australia, pers. comm. 2000). This number includes species that have been recorded outside their normal range (vagrant species), as well as species that are strict migrants, moving regularly from one country to another, often from breeding to non-breeding areas.

Because the actions of humans in other parts of the world can affect these species, several agreements have been negotiated between the Commonwealth government and the countries where the species migrate. These include the Japan and Australia Migratory Bird Agreement and the China and Australia Migratory Bird Agreement. At the multilateral level, Australia is involved in the United Nations Convention on the Conservation of Migratory Species of Wild Animals (the Bonn Convention) where they are taking the lead negotiating a regional albatross conservation agreement.

Some bird species also migrate within Australia. The Swift Parrot (*Lathamus discolor*) is an endangered bird species that only breeds in Tasmania, but migrates to the mainland for autumn-winter (March-August). This is a telling example of the threats posed to migratory species. In Tasmania, the birds are almost always associated with Blue Gum (*Eucalyptus globulus*) or Swamp Gum (*Eucalyptus ovata*) (Garnett & Crowley 2000). Both of these habitats have been reduced as a result of clearing for agriculture, residential development and intensive forestry practices. The Sharmans of Spreyton in Tasmania led community tree-planting programs to reverse the loss of habitat trees and sources of food for the Swift Parrot. On the mainland, clearance for agriculture and residential development have destroyed over 70% of the bird's habitat (Garnett & Crowley 2000). Climate change also poses a further threat to this species.

ECOSYSTEM DIVERSITY

Ecosystems: What do they Provide?

Ecosystems provide the ecological functions and processes on which consumptive and productive values depend. These functions include photosynthetic fixation, pollination, gene flow, predation, competition, maintenance of water cycles, provision of nurseries for commercial fish species (in mangroves and coral reefs in particular), regulation of climate and carbon sequestration, soil production and protection, support of symbiotic fungi essential for plant growth, storage and cycling of essential nutrients, and the absorption, breakdown and dispersal of organic wastes, pesticides, air pollutants and water pollutants, and control of crop and livestock pests through predation.

Ecosystems provide many products that do not pass through a market, termed consumptive values. The most important direct uses are food, medicine, fuel and building materials. Consumptive values are usually more diverse and depend on a much wider spectrum of the available biota than do market-based patterns of use. Many people, particularly those who live in traditional ways such as some Australian Indigenous peoples, depend directly on the natural environment for live game, firewood, edible plants, medicines, building materials, materials for weapons and transport, cultural and spiritual items, raw materials for other technology and trade goods. Natural ecosystems also provide opportunities for aesthetic, recreational and tourist use, founded on the accessibility, composition and appearance of Australian species, ecosystems and landscapes. Tourist use of natural environments is part of Australia's largest export-earning industry.

7

Increase in the Knowledge of Biodiversity

In order to sustainably manage Australian landscapes for both conservation and production purposes, there needs to be improvement in our understanding of the various elements of biodiversity. The previous sections illustrated how our understanding of even the most basic measure of biodiversity, the number of species, is poorly lacking.

TAXONOMIC ENDEAVOUR IN AUSTRALIA

Number of taxonomists and species described

Results of the ABRS survey indicate that there were around 185 taxonomists working in Australia in June 2000. These figures capture roughly 75% of the taxonomic endeavour in Australia, due to non-respondents and private collectors from Australia and other countries not covered by the survey. All Australian institutions known to hold major collections were contacted, including museums, herbaria and universities.

Taxonomists described about 2300 new species and 240 new genera between 1 July 1995 and 30 June 1999. Roughly two-thirds of the taxonomic effort was expended on animals and one-third on plants. There is currently no-one working on many of the taxa in Australia (some groups are being handled in other countries on a world basis).

Among the plants, vascular plants were the main focus. Fungi are expected to be far more numerous than vascular plants, but only about 6% of the total taxonomic effort was directed towards fungi, compared to about 75% of the effort on vascular plants. Among the animals, slightly more than one-half of the effort was directed towards the most numerous taxa (insects, arachnids and crustaceans). Just fewer than 20% of the effort was directed towards vertebrates, which make up less than 1% of all animal species.

Fungi in particular are underresourced, relative to other taxa, when the total numbers of undescribed taxa are taken into account. Fungi are important in ecosystem services and biogeochemical cycles making them just as important as vascular plants from a utilitarian perspective. The ABRS lists them as a high priority. Part of the difficulty is the lack of taxonomists in these areas and lack of people willing to work in them. Similarly, taxonomic priorities among animals reflect to some extent social and immediate economic imperatives, rather than ecological ones.

Funding for taxonomy

Taxonomic work was undertaken by 57 different institutions in all states and territories. The funding for this work comes from Commonwealth and state institutional support, and it occurs in museums and herbaria, state and Commonwealth regulatory agencies, and tertiary research and teaching institutions. Despite the significant gaps in taxonomic effort in Australia, the budget for the ABRS, which is producing the multivolumed *Flora of Australia* and *Fauna of Australia*, was cut by $400 000 in the 1999-2000 financial year. Fourteen Large Research Grants were awarded from the Australian Research Council (ARC) for taxonomic research between 1997 and 2000.

Biodiversity in Ecological Processes

Number of research programs into the role of biodiversity in ecological processes

There have been no ARC Large Research Grants into the role of biodiversity in ecological processes in the last five years.

THE USE OF SURROGATES FOR THE MANAGEMENT OF BIODIVERSITY

Number of research programs into surrogates

In the context of State of Environment reporting, and for biodiversity management in general, 'surrogates' measure the spatial distribution of biodiversity. They are distinct from indicators that measure the response of ecosystems to disturbance, and from umbrella species and flagship species that provide de facto protection for species that occupy the same habitat.

The taxon-based biodiversity surrogates approach targets resource management or landscape restoration efforts at a group of species and assumes that the needs of other taxa will be met.

A simple strategy is to conserve areas that incorporate a range of environmental factors (Faith & Walker 1993). Environmental domains are geographical regions that enclose a continuous range of physical environmental parameters that are expected to be important in determining the distributions of species.

Vegetation maps are perhaps the most frequently used biodiversity surrogates. Much of the vegetation of the Australian continent has been classified and mapped (Commonwealth of Australia 1990; Specht et al. 1995). It is assumed that protection of a proportion of each vegetation type will protect sufficient proportions of the populations of other organisms. Vegetation maps may fail as surrogates in cases where sets of species are dependent on particular successional stages within a vegetation community (e.g. the old growth stage of a particular type of forest), or when species respond to environmental variables to which the vascular flora are insensitive.

No ARC Large Research Grants have been awarded on the subject of surrogates between 1995 and 2000.

Types 1, 2 and 4 have been proposed as indicators of biodiversity and types 3, 5, 6 and 7 as indicators of abiotic conditions and/or changes in ecological processes.Taxon-based biodiversity surrogate schemes have wide appeal because it is simply impossible to measure, monitor and manage all of biodiversity (Burgman & Lindenmayer 1998). The fundamental assumption of all taxon-based surrogate schemes is that if resource management or landscape restoration efforts are targeted at a group of species, the needs of other taxa will be provided.

However, as early as the 1980s, several workers raised concerns about the conceptual, theoretical and practical basis for taxon-based surrogate schemes (e.g. Landres et al. 1988). None of these concerns have been adequately answered in the intervening years (Lindenmayer et al. 2000). Some of the many problems which afflict taxon-based surrogate schemes are outlined below.The effects of human perturbation such as landscape change and habitat fragmentation varies for each species and also between groups of species.

Hence, the response of a given species or suite of species to landscape modification may reveal very little about the response of many other species in the same or different assemblage or group.Any species that is the specific target for conservation by particular management actions can no longer be an independent yardstick of those actions and, in turn, be regarded as a suitable surrogate for other taxa.There are problems stemming simply from choosing the wrong biodiversity surrogate that can arise from a lack of understanding of the causal relationship between the response of that species and the ecosystem conditions for which it is supposed to be indicate. There are also problems stemming simply from choosing the wrong indicator species. The case of the Bivalve Mollusc (*Velesunio ambiguus*) in Australian river systems is a classic example.

Early research suggested that the species was an indicator of the presence of heavy metals (Walker 1981). Subsequent work found that the uptake of heavy metals by *Velesunio ambiguus*did not reflect the extent of pollution in the surrounding riverine system, making the mollusc an unreliable, and thus entirely unsuitable, indicator species (Millington & Walker 1983). Robust causal relationships between surrogates and other elements of biodiversity have never been demonstrated (Lindenmayer et al. 2000).A recent study of surrogate schemes by Andelman and Fagan (2000) examined the efficacy of an array of types of taxon-based surrogate schemes including indicator species, flagship species and umbrella species. Andelman and Fagan (2000) found that none of the surrogate schemes captured more species or better protected habitat than a given species selected at random from the large databases they assembled to conduct their tests.

Thus, a key problem with taxon-based surrogate schemes is that when a landscape is managed or restored in an attempt to meet the requirements of a given suite of species such as birds (e.g. through the focal species approach) it may be inappropriate to automatically assume that the food, shelter and breeding requirements of other plants and animals in the landscape have also been met.The inherent problems associated with the use of indicator species and other biodiversity surrogate schemes means that other approaches may be needed to conserve biodiversity as part of ecologically sustainable natural resource management. In the case of forest landscapes, Lindenmayer et al. (2000) recommended the adoption of what they termed 'structure-based' indicators. These included stand and landscape (spatial) level features of forests such as stand structural complexity and plant species composition, connectivity and heterogeneity. In addition to these structure-based indicators, Lindenmayer et al. (2000) advocated the following four key approaches to enhance biodiversity conservation in forests:

- The establishment of biodiversity priority areas (e.g. reserves) managed primarily for the conservation of biodiversity
- Within production forests, the application of structure-based indicators, including structural complexity, connectivity and heterogeneity
- The deployment of a risk-spreading approach in wood production forests using multiple conservation strategies at multiple spatial scales

- The adoption of an adaptive management approach to test the validity of structure-based indices of biodiversity by treating management practices as experiments.

TAXON-BASED BIODIVERSITY SURROGATES

Taxon-based biodiversity surrogates schemes have been used widely in conservation management efforts in many parts of the world. The search for indicators of biodiversity has tended to focus on biological entities (e.g. gene frequencies, populations, species, species assemblages and communities) that might function as surrogates or proxies for other forms of biodiversity and/or reflect changes in ecosystem patterns or processes (Burgman & Lindenmayer 1998). Many types of biodiversity surrogate schemes have been proposed. Some of these include: indicator species, management indicator species, keystone species, umbrella species, and the focal species approach (Lindenmayer et al. 2000). The biodiversity surrogate scheme that has received greatest attention has been 'indicator species'.

The term indicator species has been used to mean many different things. Some examples of types of indicator species include:

1. A species whose presence indicates the presence of a set of other species and whose absence indicates the lack of that entire set of species
2. A keystone species, *sensu* Terborgh (1986), which is a species whose addition to, or loss from, an ecosystem leads to major changes in abundance or occurrence of at least one other species (e.g. Mills et al. 1993)
3. A species whose presence indicates human-created abiotic conditions such as air or water pollution (often termed a pollution indicator species, Spellerberg 1994)
4. A dominant species in the sense that it provides much of the biomass or number of individuals in an area
5. A species that indicates particular environmental conditions like certain soil or rock types (Klinka et al. 1989)

6. A species thought likely to be sensitive to, and to therefore serve as an early warning indicator of, environmental changes like global warming (Parsons 1991) or modified fire regimes (Wolseley & Aguirre-Hudson 1991) (sometimes termed a bioindicator species)
7. A management indicator species, which is a species believed to reflect the effects of a disturbance regime or the efficacy of efforts to mitigate disturbance effects (Milledge et al. 1991).

THE USE OF BIOREGIONS

Proportion of bioregions covered by biological surveys

The data on this indicator are not readily available at a national or regional level, even for areas with relatively comprehensive survey records such as north-east New South Wales.

LONG-TERM MONITORING

Long-term Monitoring and Research Sites

Long-term monitoring sites are permanent sample locations set up to record trends in a range of ecological and biological characteristics. In Australia, there are numerous agencies and programs responsible for the upkeep and continued measurement of long-term monitoring sites. The first long-term monitoring site was a CSIRO site established at Gilruth Plains, Qld, in 1944. The 20 ha grazing exclosure is located within the Warrego floodplain and includes a mix of Mitchell Grass downs, Gidgee drainage line and Spinifex sandhill vegetation. A total of 14 photopoints were established in 1944 and have been rephotographed and interpreted at irregular intervals since. Several studies into the ecology and demography of Mitchell Grass (*Astrebla* spp.) have been conducted using these data, and the understanding from this study informs the management and resource use of the plains.

There is a broad range of motivations for the establishment of long-term monitoring sites. Many of the early sites were established to monitor production systems such as forests and fisheries. Most of the ecological monitoring sites with a non-production focus were established in the 1980s and 1990s, with some notable exceptions. There are also many community monitoring programs which are covered in following sections. Over 80 biodiversity or monitoring programs, covering 1995 to 2001, are also recorded for the AAT.

Global scientific interest in developing a Long-Term Ecological Research (LTER) program is expanding very rapidly, reflecting the increased appreciation of their importance in assessing and resolving complex environmental issues. In 1993, the International Long Term Ecological Research (ILTER) network was formed to develop a worldwide program, and the infrastructure necessary to facilitate communication and to manage distributed databases. Australia is a member of this international network and has four sites registered - three of these focus on production forests (two in Queensland and one in Tasmania) and the fourth, which began operations in 1998, is centred around rainforest canopy research at Cape Tribulation, north Queensland.

The only national register of long-term research and monitoring sites is maintained by the Ecological Society of Australia (ESA). The database mainly contains sites contributed by members of the Society, and would benefit from a more systematic approach to collecting information. However, with over 130 sites listed, this database represents a very useful starting point to develop a national strategy. The database has site location details for roughly 90% of the study sites listed.

Of the more than 130 long-term monitoring and research sites in the ESA database, only eight are comprehensive in the sense that they sample vertebrates, invertebrates and plants. Most sites are dedicated to sampling just one, or more rarely two, taxonomic groups. There is a distinct bias towards monitoring vascular plants, and there is a very uneven distribution of effort across taxa and among ecosystems.

8

Ecology Movements and Conflicts Over Natural Resources

The recent period in human history contrasts with all the earlier ones in its strikingly high rate of resource utilization. Ever expanding and intensifying industrial and agricultural production has generated increasing demands on the world's total stock and flow of resources. These demands are mostly generated from the industrially advanced countries of the North and the industrial enclaves in the underdeveloped countries of the South. Paradoxically, the increasing dependence of the industrialised societies on natural resources, through the rapid spread of energy and resource-intensive production technologies, has been accompanied by the spread of the myth that increased dependence on modern technologies implies a decreased dependence on nature and natural resources This myth is supported by the introduction of a long and indirect chain of resource utilisation which leaves invisible the real material resource demands of the industrial processes.

Through this combination of resource intensity at the material level and resource indifference at the conceptual and political levels, conflicts over natural resources generated by the new pattern of resource utilisation are generally shrouded and overlooked. These conflicts become visible when resource and energy-intensive industrial technologies are challenged by communities whose survival depends on the conservation of resources threatened by destruction and overexploitation, or when the devastatingly destructive potential of some industrial technologies is demonstrated as in the Bhopal disaster.

For centuries, vital natural resources like land, water and forests had been controlled and used collectively by village communities thus ensuring a sustainable use of these renewable resources. The first radical change in resource control and the emergence of major conflicts over natural resources induced by non-local factors was associated with colonial domination of this part of the world. Colonial domination systematically transformed the common vital resources into commodities for generating profits and growth of revenues. The first industrial revolution was to a large extent supported by this transformation of commons into commodities which permitted European industries access to the resources of South Asia.

With the collapse of the international colonial structure and the establishment of sovereign countries in the region, this international conflict over natural resources was expected to be reduced and replaced by resource policies guided by comprehensive national interests. However, resource use policies continued along the colonial pattern and, in the recent past, a second drastic change in resource use has been initiated to meet the international requirements and the demands of the elites in the Third World, leading to yet another acute conflict among the diverse interests. The most seriously threatened interest, in this conflict, appears to be that of the politically weak and socially disorganised group whose resource requirements are minimal and whose survival is primarily dependent directly on the products of nature outside the market system. Recent changes in resource utilisation have almost wholly by-passed the survival needs of these groups. These changes are primarily guided by the requirements of the countries of the North and of the elites of the South.

This book analyses environmental conflicts in contemporary human society. In general it relates to societies all over the world, but in particular it addresses the most intense and emerging social contradictions in India related to conflicts over natural resources.

Science and technology are central to these conflicts because while scientific knowledge has been used by contemporary societies to considerably enlarge man's access to natural resources, it has also allowed the utilisation natural resources at extremely high rates. The contemporary period is characterised by the emergence of ecology movements in all parts of the world which are attempting to redesign the pattern and extent of natural resource utilisation to ensure social equality and ecological sustainability. Ecology movements emerging from conflicts over natural resources and the people's right to survival are spreading in regions like the Indian subcontinent where most natural resources are already being utilised to fulfil the basic survival needs of a large majority of people. The introduction of resource and energy-intensive production technologies under such conditions leads to economic growth for a small minority while, at the same time, undermines the material basis for the survival of the large majority. In this way, ecology movements have questioned the validity of the dominant concepts and indicators of economic development. The ideology of economic development, which remained almost monolithic in the post World War II period, is thus faced with a major foundational challenge. In this chapter an attempt has been made to provide a systematic conceptual framework for analysing the processes and structures of modern economic development from an ecological -perspective. It attempts to analyse the relationship between economic development and conflicts over natural resources to trace the roots of ecological movements. Further, in the light of the ecological perspective, it examines the fundamental assumptions and categories of modern development economics that are used to determine the objectives of economic development as well as the criteria for the choice of technologies that are used to achieve these objectives.

ECONOMIC DEVELOPMENT AND ENVIRONMENTAL CONFLICTS IN INDIA

A characteristic of indian civilization has been its sensitivity to natural ecosystems. vital renewable natural resources like vegetation, soil and water were managed and utilised according to well defined social norms that respected the known ecological processes. The indigenous modes of natural resources utilisation were sensitive to the limits to which these resources could be used It is said that the codes of visiting important pilgrim centres Badrinath in the sensitive Himalayan ecosystem, included a maximum stay of one night so that the temple area would not put excess pressure on the local natural resources base. In the precolonial indigenous economic processes, the levels of utilisation of natural resources were not significant enough to result in drastic environmental problems. There were useful social norms for environmentally safe resource utilisation and people protested against the destructive use of resources even by kings. A major change in the utilisation of natural resources of India was introduced by the British who linked the resources of this country with the direct and large nonlocal demands of Western Europe. Natural resource utilisation by the East India Company, and later by the colonial rulers, replaced the indigenous organizations for the utilisation of natural resources, like water, forest and minerals, that were mainly managed as commons.

With the establishment of British colonial rule in India, the ever increasing resource demands of-the industrial revolution in England were largely met from colonies like India. Forced cultivation of indigo in Bengal and Bihar, cultivation of cotton in Gujarat and the Deccan led to large-scale commitment of land for the supply of raw materials for the British textile industry, the flagbearer of the industrial revolution. Forests in the sensitive mountain ecosystems like the Western Ghats or the Himalayas were felled to build battleships, or to meet the requirements of the expanding railway network. Forests of the Bengal-Bihar-Orissa region were used for running wood fuel locomotives in the early stages of railway expansion. The latter stages of colonial resource utilisation and control included the monopolization. of water rights as in the Sambhar Lake of Rajasthan or the Damoda' Canal in Bengal. Colonial intervention in natural resource management in India led to conflicts over vital renewable natural resources like water or forests and induced new forms of poverty and deprivation. Changes in resource endowments and entitlements introduced by the British came into conflict with the local people's age old rights and practices related to natural resource utilisation As a result local responses were generated through which people tried to regain and retain control over local natural resources. The indigo Movement in Eastern India, the Deccan Movement for land rights or the forest movement in all forest areas of the country, the Western

Ghats, the Central Indian Hills or the Himalayas, were obvious expressions of protest generated by these newly created conflict's. Conflicts generated by the colonial modes of natural resource exploitation could not, however, grow with a local identity. With the progress of the anti-colonial people's movement at the national level, these local protests merged with the national struggle for independence. With the collapse of colonial rule internationally, and the emergence of sovereign independent countries in the Third World like India, resolution of these conflicts at the local level became a possibility. While political independence vested the control over natural resources with the Indian state, the colonial institutional framework for natural resource management did not change in essence. Where colonialism collapsed, the slogan of economic development stepped in.

There was unfortunately no alternative institutional mechanism other than that of the classical model of development left by the British, with which the newly formed Indian state could respond to the accentuated aspirations of the Indian people for a better life. The same institutions and concepts, nurtured and developed by the colonial rulers were applied to objectives which were exactly opposite to those of the colonial period. Concepts and categories relating to economic development and natural resource utilisation that had emerged in the specific context of capitalist growth and industrialization in the centres of colonial power were raised to the level of universal assumptions and applicability. The processes which led to deprivation were now entrusted with the responsibility of basic needs satisfaction. No serious thought was given to the fact that the historical specificity of early industrial development in Western Europe necessitated the permanent occupation of the colonies and the undermining of the local 'natural economy'.' This inexorable logic of resource exploitation, exhaustion and alienation integral to the classical model of economic development based on resource intensive technologies led Gandhi to seek an alternate path of development for India when he wrote:

> God forbid that India should ever take to industrialism after the manner of the West. The economic imperialism of a single tiny island kingdom (England) is today keeping the world in chains. If an entire nation of 300 million took to similar economic exploitation, it would strip the world bare like locusts.

While Gandhi's critique was a forewarning against the problems likely to arise by following the classical path of resource-intensive development, at the time of India's independence, there was no clear and comprehensive work plan to realise the Gandhian dream of alternate development that would be resource prudent and would satisfy basic needs. The issues of resource constraints of economic development were, therefore, not highlighted at the theoretical level, partly due to the tremendous pressure of the enhanced developmental aspirations of a newly independent nation, and partly due to the lack of internalization of natural resource parameters within the framework of economics. As the scale of economic development activities escalated from one Five Year Plan to another, the disruption of ecological processes that maintain the productivity of the natural resource base started becoming increasingly apparent. The classical model of economic development in the case of the newly independent nations resulted in the growth of urban-industrial enclaves where commodity production was concentrated, as well as rapid exhaustion of the internal colonies whose resources supported the enhanced demands of these enclaves.

In the absence of ecologically enlightened resource management methods, the pressure of poverty enhanced the pace of economic development activities in the hope of a quick improvement in the standard of living for all, as in the case of Western Europe. For example, commercial forestry earned more revenue by making increasing amount of timber and pulpwood available in the market but in the process reduced the multipurpose biomass productivity or damaged the hydrology of the forests. People dependent on non-timber biomass outputs of forests like leaves, twigs, fruits, nuts, medicines and oils were unable to sustain themselves, in the face of the commercial exploitation of forests. The changed hydrological character of the forests affected both the micro-climate and the stream flows, disturbing the hydrological stability and affecting agricultural production.

There are similar examples from all parts of the country, related to almost all massive developmental interventions in India's natural resource system. Ecological degradation and economic deprivation generated by the resource insensitivity and intensity of the classical model of development have resulted in environmental conflicts, an understanding of which is imperative for the reorientation of our current development priorities and concepts. It is becoming in creasingly clear that these classical concepts and priorities are being used as an alibi to direct 'development' at the national level, while the educated minority elite is the main beneficiary of these 'development' processes.

The ecology movements that have emerged as major social movements in many parts of India are making visible many invisible externalities and pressing for their internalisation in the economic evaluation of the elite-oriented development process. In the context of a limited resource base and unlimited development aspirations, ecology movements have initiated a new political struggle for safeguarding the interests and survival of the poor, the marginalised, including women, tribals and poor peasants.

ECOLOGY MOVEMENTS AND SURVIVAL

The intensity and range of ecology movements in independent india have continuously widened as predatory exploitation of natural resources to feed the process of development has increased in extent and intensity. This process has been characterised by the massive expansion of energy and resource-intensive industrial activity and major development projects like large dams, forest exploitation, mining and energy-intensive agriculture. The resource demand of development has led to the narrowing of the natural resource base for the survival of the economically poor and powerless, either by direct transfer of resources away from basic needs or by destruction of the essential ecological process that ensure renewability of the life-supportirig natural resources.

In the light of this background, ecology movements emerged as the people's response to this new threat to their survival and as a demand for the ecological conservation of vital life-support systems. The most significant life-support systems in addition to clean air are the common property resources of water, forests and land on which the majority of the poor people of India depend for survival. It is the threat to these resources that has been the focus of ecology movements in the last few decades.

Among the various ecology movements in India, the Chipko movement (embrace the trees to oppose fellings) is the most well known. It began as a movement of the hill people in the state of Uttar Pradesh to save the forest resources from exploitation by contractors from outside.' It later evolved into an ecological movement that was aimed at the maintenance of the ecological stability of the major upland watersheds in India. Spontaneous people's response to save vital forest resources was seen in Jharkhand area in Bihar-Orissa border region as well as in Bastar area of Madbya Pradesh where there were attempts to convert the mixed natural forests into plantations of commercial tree species, to the complete detriment of the tribal people. In the southern part of India the Appiko movement, which was inspired by the success of the Chipko movement in the Himalayas, is actively involved in stopping illegal over-felling of forests and in replanting forest lands with multipurpose broad leaved tree species. In Himachal Pradesh the Chipko activists have intensified their opposition to the expansion of monoculture plantation of the commercial Chir Pine (Pinus roxburghii). In the

Aravalli Hills of Rajasthan there has been a massive programme of tree planting to give employment to those hands which were hitherto engaged in felling of trees.

The exploitation of mineral resources, in particular the opencast mining in the sensitive watersheds of the Himalayas, the Western Ghats and Central India have also resulted in a great deal of environmental damage. As a consequence, environmental movements have come up in these regions to oppose the reckless mining operations. Most successful among them is the movement against limestone quarrying in the Doon Valley. Here, volunteers of the Chipko movement have led thousands of villagers, in peaceful resistance, to oppose the reckless functioning of limestone quarries that is seen by the people as a direct threat to their economic and physical survival.'

While the Doon Valley instance has a long history of popular opposition to the quarrying of limestone and a Supreme Court order has restricted the area of quarrying to a minimum, examples of such success' of ecology movements are rare People's ecology movements against mineral exploitation in the neighbouring areas of Almora and Pithoragarh still seem to be ignored, probably due to the relative isolation of these interior areas. Beyond the Himalayas, the ecology movement in the Gandhamardan Hills in Orissa against the ecological havoc of bauxite mining has gained momentum and it draws inspiration from the Chipko movement.

The mining project of the Bharat Aluminium Company (BALCO) in the Gandhamardan Hills is being opposed by local youth organisations and tribal people whose survival is directly under threat. The peaceful demonstrators have claimed that the project could be only continued 'over our dead bodies. The situation is more or less the same in large parts of Orissa-Madhya Pradesh region where rich mineral and coal deposits are being opened up for exploitation and thousands of people in these interior areas are being pushed to deprivation and destitution. This is also true of the coal mining areas around the energy capital of the country in

Singrauli. In these interior areas of Central India, movements against both mining and forestry are becoming increasingly volatile and people's resistance is growing.

Large river valley projects, which are coming up in India at a very rapid pace, is another group of development projects against which people have organised ecology movements. The large-scale submersion of forest and agricultural lands, a prerequisite for the large river valley projects, always takes a heavy toll of dense forests and the best food growing lands. These have usually been the material basis for the survival of a large number of people in India, specially tribal people. The Silent Valley project in Kerala was opposed by the ecology movement on the ground of its being a threat, not to the survival of the people directly, but to the gene pool of the Tropical Rainforests threatened by submersion. The ecological movement against the Tehri high dam in the UP Himalaya exposes the possible threat to people living both above and below the dam site through large-scale destabilization of land by seepage and strong seismic movements that could be induced by impoundment. The Tehri Dam Opposition Committee has appealed to the Supreme Court against the proposed dam by identifying it as a threat to the survival of all people living near the river Ganga up to West Bengal. Most notable among the people's movements against dams on the issue of direct threat to survival from submersion are Bedthi lcchampalli, Bhopalpatnam, Narmada Sagar, Koel-Karo, Bodhghat, etc. In the context of the already overutilised land resources, the proper rehabilitation on a land-to-land basis of millions of people displaced through the construction of dams seems impossible. The cash compensation given instead is inadequate in all respects for providing an alternate livelihood for the majority of the displaced. Destitution is thus the first and foremost precondition for initiating large dam projects

While the process of construction of dams itself invites opposition from ecology movements, the functioning of water projects dependent on the constructed dams results in further ecological disasters and movements. People's movements against widespread water-logging, salinisation and the resulting desertification in the command areas of many dams have been registered. Among them are instances of protests against the Tawa, Kosi, Gandak, Tungabhadra, Malaprabha, Ghatprabha projects and the canal irrigated areas of Punjab and Haryana. While excess water led to ecological destruction in these cases, improper and unsustainable use of water in the arid and semi-arid regions generated ecology movements in a different way. The anti-drought and desertification movement is gaining momentum in the dry areas of Maharashtra, Karnataka, Rajasthan, Orissa, etc. Ecological water use for survival is being advocated by water based movements like Pani Chetana, Pani Panchayat, and Mukti Sangharsh. Another major movement originating from the ecological destruction of resources by growth based development is spreading all along the 7,000 km long coastline of India. It is the movement of the small fishing communities against the ecological destruction caused by mechanised fishing whose instant profit motive is destroying the coastal ecology and its long-term biological productivity in a big way.

No amount of threat to survival in India from environmental hazards can be complete without a reference to the Bhopal tragedy on 2 December 1984, in which several thousand people died and several lakhs faced serious health hazards following the leakage of poisonous Methyl Iso Cyanate from a pesticide plant of Union Carbide (India) Limited. People's movements for clean air and water are growing in all parts of the country just as ecologically irresponsible industrialization is moving deeper into the hinterland in search of new resources.

DEVELOPMENT FROM THE VIEWPOINT OF THE DISPOSSESSED

Though these ecology movements relate to issues that are geographically localised, like forests or water pollution, their reverberations are national and even global in import. This macro micro dialectic is rooted in the cognitive gaps associated with development planning and this dichotomy has been analysed politically as the result of the existence of two Indias. Every development activity invariably has a need for natural resources. In the context of limited natural resources, either limited by nonrenewability or ecological limits to renewability, the resource needs of the two Indias are bound to compete with each other. In this unequal competition the survival of the less powerful but more populous micro-economy is directly threatened. This threat may be either due to resource transfer or to ecological factors leading to resource degradation. Yet the significance of ecology movements does not merely lie in the fact that they are voices of the dispossessed who are victims of the highly unequal sharing of the costs of the development process. The positive feature of these movements lies in the manner in which they make visible the hidden externalities of development based on a particular economic ideology and reveal its inherent injustice and non-sustainability. The recognition of these inadequacies and the imperatives arising from the right to survival creates another ground and another direction for development which ensures justice with sustainability, equity with ecological stability.

Ecology movements as a trend can no longer be viewed as merely specific and particular happenings. They are an expression of the universal socio-ecological impacts of a narrowly conceived development based only on short-term commercial criteria of exploitation. The impact of ecology movements cannot be assessed merely in terms of the impact of the particular development projects they originate from. The impact, in the final analysis, is on the very fundamental categories of politics, economics, science and technology which together have created the classical paradigm of development and resource use. The emerging irreversible threat to survival arising from the development process allows a reevaluation not only of some individual projects and programmes which have been shown to be ecologically destructive, but of the very conception and paradigm of development that generates such projects. These ecology movements reveal how the resource intensive demands of current development have ecological destruction and economic deprivation built into them. They also stress that the issue is not merely one of a trade-off of costs and benefits because the cost of destruction of the conditions of life and well-being is not only a matter of money, it is a matter of life itself. The most important and universal feature of ecology movements is that they are redefining the concepts of development and economic values, of technological efficiency, of scientific rationality-they are creating a new economics for a new civilisation.

THE ECONOMISTS' SLUMBER: GROWTH AGAINST SURVIVAL

THE IDEOLOGY of the dominant pattern of development derives its driving force from a linear theory of progress, from a vision of historical evolution propounded in the eighteenth and nineteenth century Western Europe and universalised throughout the world especially in the post-war development decades. The linearity of history, presupposed in this theory of progress, created the ideology of development that equated development with economic growth, economic growth with expansion of the market economy, modernity with consumerism and non-market economies with backwardness. The diverse traditions of the world, with their distinctive technological, ecological, economic, political and cultural structures, were driven by this new ideology to converge into a homogeneous monolithic order modelled on the particular evolution of the west. The notion of development as an ideology was based on the universalisation of the western economic tradition and of equating development with economic growth alone and its unquestioned acceptance as progress.

The Rostownian model of the stages of economic growth is the clearest articulation of these assumptions. Rostow presents change as taking place in three stages. The first stage consists of traditional society: 'whose structure is developed within limited production functions, based on pre-Newtonian science and technology and on pre-Newtonian attitudes towards the physical world.... The central fact about the traditional society was that a ceiling existed on the level of attainable output per head The totality of development experiences, however, does not reflect this simple linearity and stage by stage evolution. The interrelationship between resources within the same ecosystem as well as interlinkages between economic activities, between segments of society makes the economic development process more complex and multidimensional. Viewing the world as an ecologically interrelated whole leads to a concept of development that puts a premium on maintaining the ecological balance and integrity while satisfying basic human needs.

In this context, the 'backwardness' and 'low productivity' of non-western societies is based on the assumption of the ideology of classical development that recognises productivity only in the context of commodity production. The 'high productivity' of the latter similarly has been based on a narrow and specific interpretation of productivity.

The resource intensity of modern production processes, geared towards profit maximization in the absence of the awareness of other forms of productivities, leads to ecological deterioration and loss of resource productivity, which remain hidden externalities in development economics. The internalization of such negative externalities over a large temporal and spatial horizon, in many instances, render the 'high productivity' processes extremely unproductive.

The second stage of Rostow, characteristic of the dual sector model, originates from a misleading representation of the material foundations of the visible and formal development process. In the context of a limited resource base, the resource demands of the development process are often satisfied by diverting resources away from survival needs and life-support functions. Modernisation and economic growth based on resource-intensive processes compete for the same resources and are also used for the satisfaction of basic survival needs, either directly, or through the destruction of ecological functions of these resources. The second stage is clearly not a temporary co-existence of two unrelated sectors, namely, the 'dynamic and progressive' modern and the 'stagnant and backward' traditional.

There is a distinct relationship between these two sectors in that the 'dynamism' of the modern is fuelled by a continuous and unequal resource flow from the traditional. The growth and productivity of the modern has to co-exist with the poverty and backwardness of the traditional. In the context of highly unequal sharing of the cost of economic growth, visible development accrues to the privileged while invisible underdevelopment accrues to the dispossessed. The Rostownian approach assumes that in the process of development 'the economy exploits hitherto unused resources,' which is true in the case of resource abundance. However, in the present context vital natural resources like forests, water and land are all scarce and have a number of competing requirements and demands on them.

These could be associated with the maintenance of ecological processes of renewability of natural resources or of the life-support system of those externalized by the formal process of development. The diversion of resources otherwise needed for human survival or for safeguarding the ecological processes remain invisible. Thus, in the context of the conflicting demands on scarce resources, economic growth leads to economic polarization and not necessarily to universal prosperity. Since the introduction of new technologies often leads to diversion of resources needed for survival, we have called the resultant social and economic inequalities 'technological polarisation. The rapid growth of people's ecology movements is a symptom of this polarisation and a reminder that natural resources play a vital role in the survival of people. Their diversion or destruction through other uses, therefore, leads to impoverishment and an increasing threat to survival. Underdeveloped societies are not those that are yet to be affected by growth and development, as the dual sector model supposes. The real underdevelopment of the hinterland takes place simultaneously as an integral part of the whole process of contemporary growth and development in which gains accrue to one section of society or nation and the costs, economic or ecological, are borne by the rest. From within societies and nations enjoying the advantages of resource use, Rostow's take-off stage can be seen as a reality. When one views the process of development from the perspective of those who are underdeveloped as a result of its resource intensity, the 'take-off' often gets translated into 'roll-down' into underdevelopment or ecological disasters.

Britain's 'take-off' at the end of the eighteenth century was made possible by the underdevelopment of its colonies in three continents. The destruction of the Indian textile industry and Indian agriculture, the slave trade from Africa and the genocide of the indigenous North American people were the preconditions for the economic growth of the centres of modern industry in Britain. The illusion of the contemporary take-off stage in countries like India and the vision of a flight to the twenty-first century are made possible through a similar process of the invisible destruction of the base for survival of millions of marginal people The opposition of ecology movements to resource destructive development and growth is rooted in the recognition that the creation of resources for growth is achieved through the destruction of resources for the survival of people. The Rostownian fiction of the take-off of a whole society with an improved quality of life for all its members ignores the economic polarisation and ecological destruction inherent in resource-intensive development. It appears real because under the historical conditions of colonialism or enclavised development, the invisible costs of growth are borne by the colonies or hinterlands.

The geographical separation of the regions benefiting from and the regions losing in the process left the resource destruction of the colonies and hinterlands invisible and led to the superficial impression that economic growth takes place in an absolute sense. This impression was used to universalize the Rostownian model for all countries, all people and all historical periods and this became the ideology of development. The ideological universalisation and enclavisation of the process of growth and development is the reason for the simultaneous existence of underdevelopment alongside economic growth in newly independent countries like India, which accepted rapid and resource intensive industrialisation as the path towards development. Like the erstwhile colonies, interior and resource rich areas of the country, are bearing the costs of resource diversion and destruction to run the resource-intensive process of development. As a result, communities living in these interior regions and depending on the local resources are facing a serious threat to their survival.

The ecological relationship of the growth of affluence for a few regions and some people on the one hand, and the collapse of the resource base for survival of many on the other, clearly contradicts Rostow's notion of the third stage of take-off in which 'old blocks and resistances' are overcome and the prosperity of the enclave becomes pervasive throughout society. The impoverishment of the peripheries and the erosion of the resources and rights of marginal communities actually pay for the material basis of the prosperity of the enclaves. This prosperity can neither be reproduced for regions and peoples whose impoverishment and deprivation are rooted materially and ecologically in the same process of growth, nor can the enclavisation process be sustained. The new forms of poverty and dispossession create new 'blocks and resistances' to the diffusion of the development process, making enclave development and underdevelopment of the hinterland a permanent feature of development based on resource-intensive processes.

Dichotomising tendencies and principles of exclusion seem to reflect the situation more realistically than the linear model of progress. The simplistic dichotomy between the modem and traditional sectors of the linear model is misleading because the traditional itself is transfommed and underdeveloped by the resource demands of the modern sector. This misleading dichotomy needs to be replaced by the more complex contradiction between sectors of society making conflicting and unequal demands on limited resources; between demands for profits and requirements of survival; between sustainable and non-sustainable patterns of resource use; and between the socially just and unjust use of natural resources. The reality of the ecological non-sustainability of the accepted development model and the threat to survival arising from it need to be internalised into a new framework for the understanding of economics and technology in a more realistic and less illusionary manner. Ecology movements are providing these insights for this new realism based on resource sensitivity and recognition of the people's right to survival.

While the above analysis emanates from the situation in the market economy-oriented countries of the Third World, the issues raised by it are universal in character. No doubt the anarchy of growth is most reckless in the market economy-oriented Third World countries but serious rethinking about the delicate relationship between economy and ecology is going on in both the advanced market economies and the socialist countries. As the entire world prepares to enter the third millennium humankind as a whole is feeling a special responsibility towards the global future. Human being is looking for a new philosophy to live in harmony with nature and ecology that is needed to give a new meaning and relevance to economics..

The Three Economies of Natural Resources

A new and holistic relationship between economics and ecology has to depend on a holistic understanding of the natural resource process and utilizations associated with human societies and the natural ecosystems. The dominant ideology of development. which guides development activities almost exclusively, has been classically concerned only with the use of natural resources for commodity production and capital accumulation. It ignores the resource processes that have been regenerating natural resources outside the realm of human existence. It also ignores the vast resource requirements of the large number of people whose needs are not being satisfied through the market mechanisms. The ignorance or neglect of these two vital economies of natural resources, the economy of natural processes and the survival economy, explains why ecological destruction and threat to human survival have remained hidden negative externalities of the development process. To make good for this shortcoming it is necessary to comprehend the place of natural resources in all the three economies.

NATURAL RESOURCES IN THE MARKET ECONOMY

The incompetence of modern economics in dealing with natural resources in their ecological totality has been voiced by many. The most penetrating description, however, comes from Georgescu Roogen who wrote:

The no deposit no return analogy benefits the businessman's view of economic life. For, if one looks only at money, all one can see is that money just passes from one hand to another: except by regrettable accident it never gets out of the economic process. Perhaps the absence of any difficulty in securing raw materials by those countries where modern economics grew and flourished was yet another reason for economists to remain blind to this crucial economic factor. Not even the wars the same nations fought for the control of the world's natural resources awoke the economists from their slumberer

While trade and exchange of goods and services have always characterized human societies, the elevation of the market to the position of the highest organising principle of society led to the neglect of the other two vital economies in development thought. The hidden negative externalities of the development processes governed by the principles of the market have, thus, created new forms of poverty and underdevelopment. Various case studies described in this volume will substantiate such a claim.

The major problem is that when exclusive attention is being given to monetary flows, requirements of natural resources not backed up by suitable purchasing power cannot be registered on the economic scene. As a result, specially in the context of Third World countries, the place of natural resources in the economy of natural resource production (or nature's economy) or in the survival economy of non-market consumption for the biological sustenance of the marginalised poor gets completely ignored. The political economy of ecology movements cannot be understood without a clear comprehension of the place of natural resources in the three distinct economies.

Ecology movements are the first indicators of compatibility and conflict among the three competing demands on natural resources. In this way the articulation of these three economies provides the foundation of a framework for an ecologically sustainable and equitable process of economic development that ensures survival and does not threaten it. The benefits and costs associated with development projects thus need to be evaluated not only in terms of the framework of the market economy but also in terms of the other two economies associated with natural resources.

THE ECONOMY OF NATURAL ECOLOGICAL PROCESSES

The terms ecology and economy are rooted in the same Greek word 'oikos' or household. Yet in the context of market-oriented development they have been rendered contradictory: 'Ecological destruction is an obvious cost for economic development'-a statement which is often repeated to ecology movements. Natural resources are produced and reproduced through a complex network of ecological processes. Production is an integral part of this economy of natural ecological processes but the concepts of production and productivity in the context of development economics have been exclusively identified with the industrial production system for the market economy. Organic productivity in forestry or agriculture has also been viewed narrowly through the production of marketable products of the total productive process. This has resulted in vast areas of resource productivity, like the production of humus by forests, or regeneration of water resources, natural evolution of genetic products, erosional production of soil fertility from parent rocks, remaining beyond the scope of economics.

Many of these productive processes are dependent on a number of ecological processes. These processes are not known fully even within the natural science disciplines and economists have to make tremendous efforts to internalize them. Paradoxically, through the resource ignorant intervention of economic development at its present scale, the whole natural resource system of our planet is under threat of a serious loss of productivity in the economy of natural processes. At present ecology movements are the sole voice to stress the economic value of these natural processes.

The market-oriented development process can destroy the economy of natural processes by over exploitation of resources or by the destruction of ecological processes that are not comprehended by economic development. And these impacts are not necessarily manifested within the period of the development projects. The positive contribution of economic growth from such development may prove totally inadequate to balance the invisible or delayed negative externalities stemming from damage to the economy of natural ecological processes. In the larger context, economic growth can thus, itself become the source of underdevelopment. The ecological destruction associated with uncontrolled exploitation of natural resources for commercial gains is a symptom of the conflict between the ways of generating material wealth in the economies of-market and the natural processes. In the words of Commoner: 'Human beings have broken out of the circle of life driven not by biological needs, but the social organisation which they have devised to 'conquer' nature: means of gaining wealth which conflict with those which govern nature."

The Survival Economy

Modern economics and the concept of development cover a miniscule portion in the history of economic production by human beings. The survival economy has given human societies the material basis of survival by deriving livelihoods directly from nature through self-provisioning mechanisms. In most Third World countries large numbers of people are deriving their sustenance in the survival economy in ways that remain invisible to market oriented development. Within the context of a limited resource base the destruction of the survival economy takes place through the diversion of natural resources from directly sustaining human existence to generating growth in the market economy. Sustenance and basic needs satisfaction is the organising principle for natural resource use in the survival economy whereas profits and capital accumulation are the organising principles for the exploitation of resources for the market. Human survival in India even today is largely dependent on the direct utilisation of common natural resources."

Ecology movements are voicing their opposition to the destruction of these vital commons so essential for human survival. Without clean water, fertile soils, and crop and plant genetic diversity economic development will become impossible. Sometimes by omission and sometimes by commission formal economic development activities have impaired the productivity of common natural resources which has enhanced the contradiction between the economy of natural processes and the survival economy.

The organising principles of economic development based on economic growth renders valueless all resources and resource processes that are not priced in the market and are not inputs to commodity production. This premise very often generates economic development programmes that divert or destroy the resource base for survival. While the diversion of resources, like diversion of land from multipurpose community forests to monoculture plantations of industrial tree species, or the destruction of common resources, or the diversion of water from staple food crops and drinking water needs to cash crops are frequently proposed as programmes for economic development in the context of the market economy, they create economic underdevelopment in the economies of nature and survival. Ecology movements are aimed at opposing these threats to survival from market based economic development. Thus in the Third World, ecology movements are not the luxury of the rich; they are a survival imperative for the majority of people whose survival is not taken care of by the market economy but is threatened by its expansion.

The political foundation of ecology movements lies in their capacity to enlarge the spatial, temporal and social bases for the evaluation of economic development projects-in their capacity to bring into the picture all the three economies described earlier. A new economics of development will emerge only when these three economies can be conceptualized within a single framework.

Technology Choice Towards Holistic Ecological Criteria

When economic development programmes are viewed from the perspective of all the three economies, a clearer view of the political economy of conflicts over natural resources is expected to emerge. In the dominant mode of economic development, perceived within the framework of the market economy, mediation of technology is assumed to lead to the control of larger and larger quantities of natural resources, thus turning scarcity into abundance and poverty into affluence: Technology, accordingly is viewed as the motive force for development and the vital instrument that guarantees freedom from dependence on nature ' The affluence of the industrialized west is assumed to be associated exclusively with this capacity of modern technology to generate wealth.

The concept of technology per se as a source of abundance and freedom from nature's ecological limits are based in part on the limitations of the market economy in understanding in a holistic manner, the same resources which it exploits. Only when development processes are viewed in the holistic perspective of all the three economies can the scarcities and underdevelopment associated with abundance and development be clearly seen. Most resource-intensive technologies operate in the enclaves with enormous amounts of various resources coming from diverse ecosystems which are normally far away. This long, indirect and spatially distributed process of resource transfer made possible by energy-intensive long distance transportation, leaves invisible the real material demands of the technological processes of development.

The spatial separation of resource exhaustion and the creation of products have also considerably shielded the inequality creating tendencies of modern technologies. Further, it is simply assumed that the benefits of economic development based on these modern technologies will automatically percolate to the poor and the needy and growth will ultimately take care of the problems of distributive justice. This would, of course, be the case, if growth and surplus were in a sense absolute and purchasing power existed in all socio-economic groups. None, however, is correct. Surplus is often generated at the cost of the ecological productivity of natural resources or at the cost of exhausting the capital of non-renewable resources. For the poor, the only impact of such economic activity often is the loss of their resource base for survival.

It is thus no accident that modern, efficient and 'productive' technologies 'creased within the context of growth in market economic terms are associated with heavy social and ecological costs. The resource and energy intensity of the production processes they give rise to demands ever increasing resource withdrawals from the natural ecosystems. These excessive withdrawals in the course of time disrupt essential ecological processes and result in the conversion of renewable resources into non-renewable ones. Over time, a forest provides inexhaustible supplies of water and biomass including wood, if its capital stock, diversity and hydrological stability are maintained and it. is harvested on a sustained yield basis. The heavy and uncontrolled market demand for industrial and commercial wood, however, requires continuous over-felling of trees which destroys the regenerative capacity of the forest ecosystems and over time converts these forests into non-renewable resources.

Sometimes the damage to nature's intrinsic regenerative capacity is impaired not directly by over-exploitation of a particular resource but indirectly by damage caused to other natural resources related through ecological processes. Thus under tropical monsoon conditions, over-felling of trees in catchment areas of streams and rivers not only destroys forest resources, but also stable, renewable sources of water. Resource-intensive industries do not merely disrupt essential ecological processes by their excessive demands for raw materials; they also destroy and disrupt vital ecological processes by polluting essential resources like air and water. In the words of Rothman: 'the private economic rationality of the profit seeking business enterprise is a murderous providence because it cannot guarantee the optimum use of resources for society as a whole. It cannot avoid continually creating situations which cause the pollution of an environment

In the context of resource scarcity where most resources are already being utilised for the satisfaction of survival needs, further diversion of resources to new uses is likely to threaten survival and generate conflicts between the demands of economic growth and the requirements of survival. It, therefore, becomes essential to evaluate the role of new technologies in economic development on the basis of their resource demands and conflict with the demands of survival. The productivity of 8 technology in the perspective of human survival must distinguish outputs in terms of their potential for satisfaction of vital or non-vital needs, because on the continued satisfaction of vital needs depends human survival. As Georgescu-Roegen points out,

> There can be no doubt about it. Any use of the natural resources for the satisfaction of non-vital needs means a smaller quantity of life in the future. If we understand well the problem, the best use of our iron resources is to produce plows or harrows as they are needed, not Rolls Royces, not even agricultural tractors.

In the context of the market economy, the indicators of technological efficiency and productivity are totally independent of the difference between the satisfaction of basic needs and luxury requirements. between resources extracted by ecologically sensitive or insensitive technologies or of the nature of the contribution of economic growth to diverse socio-economic categories. In the context of a highly non-uniform distribution of purchasing power and scanty knowledge of or respect for ecological processes, economic growth depends on production and consumption of nonvital products. The expansion of the formal sector of the economy for the production of non-vital goods often leads to further diversion of vital natural resources. For example, water-intensive production of flowers or fruits for the lucrative export market often results in water scarcity in low rainfall areas. In a world with a limited and shrinking resource base, and in the economic framework of a market economy, non-vital luxury needs are fulfilled at the cost of vital survival needs. The high powered pull of the purchasing capacity of the rich of the world can draw out necessary resources in spite of resource scarcity and resulting conflicts.

This complete lack of recognition of the resource needs of the survival economy nature's economy in the current paradigm of development economics shrouds the political issues arising from resource transfer and ecological destruction. For the economic sector based on 'efficient modern technologies', this provides an ideological weapon for increased control of the sponsors of economic development over the total natural resource endowments of the countries concerned.

The ideological and limited concept of 'productivity' of technologies has been universalised with the consequence that all other costs of the economic process become invisible. The invisible forces which contribute to the increased 'productivity' of a modern farmer or factory worker emanate from the increased consumption of non-renewable natural resources. Lovins has described this as the amount of 'slave' labour at present at work in the world. According to him, each person on earth, on an average, possesses the equivalent of about fifty slaves, each working forty hours a week. Man's annual global energy conversion from all sources (wood, fossil fuel, hydroelectric power, nuclear) at present is approximately 8 x 10 (12) watts. This is more than twenty times the energy content of the food necessary to feed the present world population at the FAO standard per capita requirement of 3,600 cals per day.

In terms of workforce, therefore, the population of the earth is not 4 billion but about 200 billion, the important point being that about 98 per cent of them do not eat conventional food. The inequalities in the distribution of this 'slave' labour between different countries is enormous, the average inhabitant of the USA, for example, having 250 times as many 'slaves' as the 'average Nigerian'. And this, substantially is the reason for the difference in efficiency between the American and Nigerian economies: it is not due to the differences in the average 'efficiency' of the people themselves. There seems no way of discovering the relative efficiencies of Americans and Nigerians: If Americans were short of 249 of every 250'slaves' they possess, who can say how 'efficient' they would prove themselves to be.

The increase in the levels of resource consumption is taken universally as an indicator of economic development. If the present level of resource consumption in the USA is accepted as the development objectives of India, the total resource demands of 'developed' India can be calculated by multiplying the current resource consumption by a factor of 250. Neither our forests nor our fields or rivers can sustain such a 'development'. When per capita resource consumption is considered, the Malthusian argument relating population with resource scarcity does not hold good. More significant than the population factor is the total resource factor. Thus, although many countries of the South have a much larger population than those of the North, the industrialized of the world consumes more grain than all the other three-quarters put together. This high consumption is due to the fact that intensive livestock production in industrialized countries accounts for 67 per cent of their total grain consumption.

This Efficient' process of livestock management for the production of meat, as reported by Odium requires 10 calories of energy input to produce one calorie of food energy. The energy subsidy provided by the capital stock of the earth's non-renewable resources makes a resource inefficient process appear as efficient in the market economy. It is interesting to note that even in the West, nearly a century ago one calorie of food was produced by using a fraction of a calorie of energy input. The same is true in the economics of water resources use in modern agriculture. When the production of high yielding varieties of seeds is evaluated, not on productivity per unit land (tons/ha) but per unit volume of water input (tons/le lit), these miracle seeds of the Green Revolution are seen as two to three times less efficient in food production than, say, the millets. The results of evaluation of the technological efficiency of processes associated with economic development, when reexamined on a holistic basis and optimised against all resource inputs, would generally lead to the conclusion that: 'the much talked of efficiency of widely practiced high technology is not intrinsically true. They are, in fact, highly wasteful of materials and pollutive (that is, destructive to the productive potential of the environment)'.

New technologies in the market economy are innovated for profit maximization and not to encourage resource prudence per se. The extent of inefficiency in the utilisation of natural resources with production processes based on resource-intensive technologies, can be illustrated with the production of soda ash, an important industrial material. In the Solvay process for the production of soda ash. the two materials used are sodium chloride and limestone.

The entire limestone used in the process ends up as waste material, 25 per cent of the sodium chloride is lost as unreacted salt. From the balance 75-80 per cent, the acidic half is lost and only the basic half goes into the final product. Therefore only 40 per cent of the raw materials consumed are actually utilised. The waste products pollute land and water resources systems. The economy of the process is artificially made good by concessions in procuring limestone, salt and fuel and further concessions in respect of land, transport, etc. It is these subsidies for natural resources which make the counter-productive processes appear efficient.

Referring to the technology of production of frozen orange juice Schnaiberg made the following remarks:

What is true of the unobtrusive shift from fresh oranges to frozen orange juice is typical of most transitions from traditional to late industrial technologies. The majority of these become more energy intensive: the energy content of all the necessary production processes increases per unit produced.... The hall mark of modern technology is its typical labour saving quality-not its energy saving aspect."

Guided by a narrow and distorted concept of efficiency and supported by all types of subsidies, technological change in market economy-oriented development continues in the direction of resource intensity, labour displacement and ecological destruction. The long-term continuation of such processes will lead to the destruction of the resource base of the survival economy and to human labour being rendered dispensable in the production processes of the market economy. The partisan assumptions of modern economic development which cannot internalise the economy of natural processes and the survival economy are thus being raised to the level of universality. As a result, with the expansion of economic development in Third World countries, the resource-intensive and socially partial development is leading to social instability and conflicts. While ecology movements in the industrially advanced countries are directed against more recent threats to survival like pollution, ecology movements in Third World countries have a much longer history related to resource exhaustion and ecological degradation of natural ecosystems. It is in these countries that the holistic ecological criteria for technology choice is needed most urgently.

The process of transformation and utilisation of natural resources for the satisfaction of societal needs determines the economic organization of human societies. At various stages of development, the dominant patterns of utilisation of natural resources have been guided by the dominant pattern of scientific knowledge, and through the generation and use of technologies that actually bridge the gap between natural resources and human needs and requirements.

A special characteristic of human societies is that they can make deliberate choices between different ways of using resources and satisfying needs. The existence of plurality of alternatives in resource use for economic development creates the need for a selection criteria to make rational decisions about the use of natural resources and technological change. A dialectical relationship exists between the criteria of technology choice and the nature of science and technology developed in response to the criteria. Traditional societies as well as modern scientific-industrial societies have adopted different systems of science and technology which differ primarily in the criteria of choice or rationality that guides resource use patterns for human needs satisfaction. The characterization of certain societies as primitive and unscientific is, thus, sociologically and epistemologically unfounded. The fact that values and rationality criteria of one form of social organization generate a particular type of science and technology matched to a particular criteria of scientificity does not imply that other social organisations lack a scientific basis for their economic activities.

If sustainable utilisation is the objective that guides the criteria of choice for a development strategy, a resource prudent technological path (T.) is rationally chosen. If maximization of the growth of man-made processes and increasing the productivity of labour is the objective, then a more resource-intensive path (T2) which is the integration of a large number of smaller technologies (t1) and in which increased resource and energy inputs allows the increase in labour productivity, is rationally chosen. In this process a large amount of secondary resources (R2-R6) are additionally required.

Traditional societies in all their diversity have, in general, shared a common set of characteristics. They have used natural resources prudently to satisfy minimum needs sustainably over centuries. Such resource use was based on

Resource flow in resource prudent t1 and resource.-intensive t2 technology chains

1. A knowledge system with an ecological understanding of nature.
2. A technological system for processing resources to satisfy human needs with minimum resource waste.
3. Rationality criteria for demarcating vital and non-vital needs and between resource destructive and resource enhancing technologies.

Traditional world views and practices deterred over-exploitation of natural resources at all levels. As they were based on ecological perceptions of nature and guided by restraints in resource use, they used technologies which prevented ecological disruption. Modernisation of traditional societies in its present form has, by and large, been taken as synonymous with the substitution of indigenous science and technology systems by the modern western system. In this manner the resource-intensive western pattern of resource use is thrust on non-western societies through modernisation.

Modern western scientific knowledge, however, differs from indigenous knowledge systems in three important ways:

1. Modern western scientific knowledge is reductionist and fragmented.
2. Modern western technological systems are based on reductionist science and are generally more resource-intensive.
3. There are no criteria of rationality or technology choices to evaluate modern science and technology on the basis of resource use efficiency or need satisfaction capability.

These characteristics of modern western science and technology systems breaks the chain, beginning with natural resources and ending in the satisfaction of human needs and demands, into small fragments of individually identifiable economic activities. This provides justification for the resource intensity of the dominant paradigm of economic development and technological change, and thus leads to ecological instabilities. Ecological crises are thus inevitable products of economic activities which are propelled towards longer and more complex and resource-intensive technological chains (T2) for the satisfaction of older needs (N.). Only individual segments(l) of the whole technological chain are examined from the narrow criteria of labour productivity. The situation is best exemplified in the case of food production. While indigenous and traditional food production practices used about half a calorie of energy to produce 1 calorie of food, the present mechanised and chemical farming techniques use 10 calories of energy to produce 1 calorie of food.

These characteristics of contemporary scientific industrial development are the primary causes for the contemporary ecological crises. The combination of ecologically disruptive scientific and technological modes, and the absence of rationality criteria for evaluating scientific and technological systems in terms of resource use efficiency, has created conditions where society is increasingly propelled towards ecological instability and has no rational and organised response to arrest and curtail these destructive tendencies.

Ecology and the Politics of Knowledge

The paradigm of modern science has evolved in the last few centuries in an environment where all economic activities were aimed at maximising the productivity of man-made processes in individual sectors of the economy. This led to the development of modern technologies with highly negative externalities which remained invisible within the conceptual framework of modern science and economics. This shortcoming emanates from three basic fallacies of modern scientific knowledge:

1. It identifies development merely with sectoral growth, ignoring the underdevelopment introduced in related sectors through negative externalities and the related undermining of the productivity of the ecosystem.
2. It identifies economic value merely with exchange value of marketable resources. ignoring use values of more vital resources and ecological processes.
3. It identifies utilization merely with extraction. ignoring the productive and economic functions of conserved resources.

Development planning based on these false identifications tends to create severe ecological problems because of its inability to recognise ecosystem linkages and the ecological processes operative in the natural world. The ecological relationships between the sectors of natural resources contribute to essential ecological processes which are frequently found to be vital for human survival. Thus, the stability of ecological processes is not merely a matter of aesthetics. An incomplete understanding of the material and economic values of ecological processes leads to the destruction of the material conditions for economic development and eventually survival

Since the availability of essential and vital resources for survival is dependent on the maintenance of essential ecological processes, economic activities which generate sectoral growth in the shortterm by destroying the essential ecological processes cannot lead to development in the long run. On the contrary, by decreasing the productivity and availability of vital resources, they initiate the process of underdevelopment. When the natural world is viewed ecologically as a system of interrelated resources which maintain the material basis for human sustenance, economic values can no longer be perceived merely as exchange values in the market. Economic values in the ecological perspective are not always equivalent to their exchange value in the market, evaluated without any significance to their use value.

As a corollary, natural resources can have economic utility that cannot be quantified through the exchange value in the market. Such economic utility includes the maintenance of essential ecological processes that support human survival and, thus, all economic activities. The economic utilisation of resources through extraction may, under certain conditions, undermine and destroy vital ecological processes leading to heavy but hidden diseconomies. The nature of these diseconomies can be understood only through the understanding of ecological processes operating in nature.

The economics of sustenance and basic needs satisfaction is, therefore, linked with ecological perceptions of nature. The economics of sectoral growth on the other hand is related to reductionist science and resource wasteful technologies which are productive in the narrow context of sectoral and labour inputs, but may be counter-productive in the context of the overall economic base of natural resources. The case studies in the following chapters are only representative of thousands of such cases seen everywhere. They reveal a certain pattern of contemporary economic development which can be identified thus:

1. Development has been equated only with the growth of manufacture in individual industrial sectors and with the increase in productivity of only man-made processes.

2. This sectoral growth of man-made processes has also led to ecological destruction of the natural resource base, affecting negatively other sectors of the economic system. This leads to the decay of systems productivity of all productive processes, man-made and natural.

As a result of this limitation of contemporary economics, economic development has, consequently, been taken to be synonymous with growth. The higher the rate of sectoral growth, the higher is the index of economic development. Possible ecological destruction caused by the resource intensity of sectoral growth that is guided purely by non-ecological economic considerations, has never been introduced in the processes of planning for economic development. The benefit-cost analysis of development projects has thus externalized those ecological changes and is incomplete in three important ways:

1. It deals with benefits and costs as profits and losses in financial ferms.
2. It deals with benefits and costs only in the narrow sectoral perspective and ignores costs generated by inter-sectoral linkages.
3. It deals with benefits that are largely available to more visible and economically powerful groups and ignores costs that are borne by the less visible and economically weaker groups. These costs and the associated underdevelopment are thus made invisible in modern economic analysis.

The utilisation and management of natural resources in India has so far been guided by the narrow and sectoral concept of productivity and restricted benefit-cost analysis. This narrow concept of productivity and benefit-cost analysis has blocked the conceptualization of the criteria of rationality of technology choice which maximizes needs satisfaction while minimising resource use, thus maximising systems productivity. For example, the clear felling of natural forests in the catchments of rivers, and planting of industrial species of trees has been justified on the grounds of increasing productivity of forests. This concept of productivity is, however, only related to productivity of industrial timber, while forests produce other forms of biomass, like fodder and green mulch, or maintain productivity of soil and water resources.

The direct impact of the clear felling of catchment forests on agricultural production through its destructive impact on soil and destabilisation of the hydrological balance is not taken into account in the calculations of the benefits and costs associated with forests. Regular floods and droughts, which are the consequences of irrational land and water management, are branded as natural disasters for which the whole nation pays heavily. Consequently, the poor and marginal groups which depend on agriculture for their livelihood face increasing impoverishment and poverty. This thrusting of negative externalities on the poor and marginal groups directly leads to the polarisation of society into two groups. One group gains from the process of narrow sectoral growth, while the poor and marginalised majority suffer because of the ecological destruction of natural resources on which they depend for survival.

The dialectical contradiction between the role of natural resources in production processes to generate growth and profits and their role in natural processes to generate stability is made visible by movements based on the politics of ecology. These movements reveal that the perception, knowledge and value of natural resources vary for different interest groups in society. The politics of ecology is thus intimately linked with the politics of knowledge. For subsistence farmers and forest dwellers a forest has the basic economic function of soil and water conservation, energy and food supplies, etc. For industries the same forest has only the function of being a mine of raw materials.

These conflicting uses of natural resources, based on their diverse functions, are dialectically related to conflicting perceptions and knowledge about natural resources. The knowledge of forestry developed by forest dwelling communities therefore evolves in response to the economic functions valued by them. In contrast, the knowledge of forestry developed by forest bureaucracies, which respond largely to industrial requirements, will be predominantly guided by the economic value of maximising raw material production. The way nature is perceived is therefore related to the pattern of utilisation of resources. Modern scientific disciplines which provide the currently dominant perspectives of nature have generally been viewed es 'objective', 'neutral' and 'universally valid'. These disciplines are, however, particular responses to particular economic interests. This economic determination influences the content and structure of knowledge about natural resources which, in turn, reinforces particular forms of resource utilisation The economic and political values of resource use are thus built into the structure of natural science knowledge.

PARTISAN SCIENCE VERSUS PUBLIC INTEREST SCIENCE

When the dominant resource use is guided by vested interests or special interest objectives, it generates a partisan science which tends to be reductionist in character. Two central assumptions underlie this reductionist perception of nature: (a) natural resources are isolated and non-interacting collections of individual resources, and (b) natural resources acquire economic value only when commercially exploited. This approach to nature is reductionist on two counts. First, it reduces nature to its constituent parts, and takes no cognizance of the relationships between the parts, and the structure and functions of the whole system. Second, it reduces economic value to a man-made construct-something which is the product of technology and capital inputs for the market. Nature's work and the work of women or marginal communities which depend on nature s productivity is thus ignored and destroyed.

Partisan science tends to be epistemologically reductionist because maximisation of special, vested interest objectives focuses on single resource functions. Partisan science must be narrowly conceived-it is inherent in its logic to perceive nature in a way that maximises the special interest objective and to be blind to ecological and environmental costs that this perception entails. Environment movements that emerge as a protest against the violation of public interest through special interest groups must therefore not merely indicate the social and environmental consequences of narrow profit maximisation. A deep and sustained resolution of such conflicts in favour of the larger public interest must be based on the emergence of a different approach to nature in the creation of a public interest science. The characteristics of this public interest science are such that it must be ecological which means: (a) it must be based on the recognition of relationships and interdependence among the various material components of nature; (b) it must be able to see and assess nature's work and assign a value to it; and (c) relatedly, it must be able to locate how nature's processes support survival, not merely profitability.

Ecology provides an epistemological framework within which alternatives to reductionist science and technology are not merely possible, but preferable too, because reductionism fails to provide faithful accounts of nature. This cognitive failure of reductionist sciences stems from the incapability of reductionism to take into account properties that emerge from relationships in nature. In this sense, the ecological foundations of an alternative science and technology differ from philosophies based on epistemological relativism. While epistemological relativism also includes the possibility of alternatives, it denies the existence of materialistic criteria of the rational choice of alternatives.

This is the limitation of the Kuhnian model, as well as other models arguing for plurality from a purely sociological or physiological perspective, not from materialist foundations. The ecological foundations for an alternative science and technology provide a materialist epistemology for evaluating the rationality of knowledge claims on the basis of their materialist adequacy in guiding action in the real and complex world. The rejection of the reductionist interpretation of materialism need not amount to an adoption of a materially vacuous philosophical position. The ecological perspective provides such a materialist alternative to reductionism. The distinction between reductionist materialism and ecology is the difference between mechanical materialism and dialectical materialism repeatedly articulated by Marx. Engels's analysis of this distinction in his critique of Duhring reads exactly like a contemporary ecological critique of reductionist science.

The analysis of nature into its constituent parts were the funda mental conditions for the gigantic strides made in our knowledge of nature during the last four hundred years. But this method of investigation has also left us a legacy of the habit of observing natural objects and natural processes in their isolation, detached from the vast interconnection of things, and therefore not in their motion, but in their repose, not as essentially changing, but as fixed constants, not in their life but in their death, in contemplating their existence it forgets their coming into being and passing away, in looking at them at rest it leaves their motion out of account because it cannot see the wood for the trees. Dialectics grasps things and their concatenation, their motion, their coming into being and passing out of existence. In a world where relationships are an actuality, the denial of such relationships and the multidimensional properties they give rise to, has created a reductionist world view and knowledge system which is inadequate in functioning in the real world. The materialist criteria provided by ecology allow for the perception of such a failure of knowledge systems through the ecological instabilities induced by them. Reductionist knowledge leads to unreliable claims about natural systems and processes on the basis of the ecological criteria of.materialist adequacy. The cognitive failure of reductionism is due to the fact that reductionist science has created ecological instabilities which in turn threaten survival. In a materialist epistemology, systems of knowledge are simultaneously systems of action.

Reductionist science leads to the human transformation of nature which is successful in creating artefacts and generating exchange value, but which fails to maintain the essential life-support systems on which human survival depends. Reductionism is not an epistemological accident. It is a particular response to an economic need of a particular form of economic organization. The reductionist world view, the industrial revolution and capitalist organisation are the philosophical, technological and economic components of the same process.

Economic growth, the achievement of this economic organisation, is materially based on externalizing the real costs of production, and on commercialising hitherto common resources to provide inputs to the production process. This entails a large withdrawal of industrial resources from the ecosystem in accordance with the demands of the market, not in accordance with the renewal capacity of resources or the needs of the people. Since it is the individual resource which generates exchange value through extraction, scientific knowledge of natural resources which is created as a response to this economic system must necessarily be reductionist. Properties of resources which stabilise ecological processes but are commercially valueless because they cannot be exchanged in the market place are ignored, and eventually destroyed. Profits and commercial exploitation shape the context in which properties of the natural systems will be perceived and known.

Scientific knowledge is not universal, objective and neutral as it is posited to be. It is always a particular response to a particular interest. When the interest is the commercial utilisation of resources for maximising exchange value, the type of knowledge system that is created is reductionist. Internalisation of profits and externalisation of costs is a normal consequence when nature is treated as if its individual components are isolated and unrelated, and the only components with economic value are those that can be transformed into commodities. The basic terms, concepts and definitions have built into them the economic values of the interest to which the knowledge is a response. In contrast, when the interest is sustainable livelihood of the people and the satisfaction of basic needs, ecological knowledge is the response.

Ecology as a public interest science is central to a just resolution of environmental conflicts in the contemporary setting because it is science and not politics that is used as the explicit justification and legitimisation of destruction, in the name of progress. 'Science' is used as a final arbiter in all resource conflicts.

The term 'scientific' is viewed as synonymous to public interest. However, since dominant science is partisan, decisions based on it will serve the special interest groups. Public interest science is a tool which makes explicit the political nature of partisan science and makes it a factor located within environmental conflicts, not a source of independent and neutral judgements about conflicts. Public interest science, however, does not merely have a critical role in the politics of knowledge and politics of the environment. It also has a constructive role in generating new paradigms of science and development based on ecological principles which ensure sustainability and justice. Probably the exemplar of public interest science is Rachel Carson's Silent Spring (1962) which exposed the destruction caused by the use of poisons in pest control and laid the foundation for alternative non-chemical means of control.

Introduction because in substance it was ecological, and in form it was different from the work of entomologists which while being critical of pesticides had remained confined to debates among entomologists and had failed to inform public debate and public policy. Silent Spring as public interest science, as a technical critique of dominant partisan science supporting pesticides, helped the growth of the environment movement since the sixties. It strengthened the movement because while exposing the cognitive weakness of partisan entomology it gave congnitive strength to ecological pest control and to the right of the individual citizen not to be poisoned by another person's pesticides. Knowledge is power, and environment movements need the cognitive power derived from public interest science. The Narmada groups have shown how this power can be used to control powerful agencies like the World Bank. Doctors in the voluntary sector working in Bhopal have shown the power of public interest science in the rehabilitation and relief provided to the victims of the gas tragedy.

Bibliography

Ausubel, F.M.: *Current Protocols in Molecular Biology*, New York, John Wiley and Sons, 1989.

Baker, Kenneth Frank: *Ecology of Soil-bourne Plant Pathogens, Prelude to Biological Control*, Berkeley, University of California Press, 1965.

Costanza, R.: *Ecological Economics*, Columbia Uni. Press., New York, 1991.

Daubenmire, F.: *Plants and Environment*, New York, Wiley, 1947.

F. Reif: *Fundamentals of Statistical Mechanics,* McGraw-Hill, Singapore, 1965.

Jameson, J. L.: *Principles of Molecular Medicine*, Totowa, Humana Press, 1998.

Jameson, J. L.: *Principles of Taxonomy*, Totowa, Humana Press, 1998.

Lynch, J.M., and Hobbie, J.E.: *Micro-organisms in Action: Concepts and Applications in Microbial Ecology*, Blackwell Scientific Publications, Oxford, 1988.

Maarten, C.: *Plants, Genes and Biodiversity*, Sudbury MA, Jones and Barlett Publishers, 2003.

Metting, Jr., F.B.: *Soil Microbial Ecology: Applications in Agricultural and Taxonomy*, Marcel Dekker, New York, 1992.

Rietman, Ed.: *Molecular Engineering of Nanosystems,* New York, Springer, 2001.

Switzer, R.L.: *Experimental Biodiversity*, New York, W.H. Freeman and Company, 1977.

Wood, E.J. Ferguson: *Marine Microbial Ecology*, London, Chapman and Hall, 1965.

Index